INVENTORS and INVENTIONS

WITHDRAWN

Volume 5

Russell, James T.— Indexes

 **Marshall Cavendish**
Reference
New York

Marshall Cavendish
99 White Plains Road
Tarrytown, New York 10591-9001

www.marshallcavendish.com

Library of Congress Cataloging-in-Publication Data

Inventors and inventions.
        p. cm.
Includes bibliographical references and index.
ISBN 978-0-7614-7761-7 (set)—ISBN 978-0-7614-7763-1 (v. 1)— ISBN 978-0-7614-7764-8 (v. 2)—
ISBN 978-0-7614-7766-2 (v. 3)—ISBN 978-0-7614-7767-9 (v. 4)—ISBN 978-0-7614-7768-6 (v. 5)   1.
Inventions—History. 2. Technology—History. 3. Inventors—History.
I. Marshall Cavendish Corporation. II. Title.

T15.I62 2007
609--dc22

                                                                    2007060868

Printed in Malaysia.

11 10 09 08 07 1 2 3 4 5

Consulting Editor: Doris Simonis, Kent State University

Contributors: Richard Beatty; Jonathan Dore; Laura Lambert; Paul Schellinger; Mary Sisson; Gwendolyn
                        Wells; Chris Woodford

MARSHALL CAVENDISH
Editor: Evelyn Ngeow
Publisher: Paul Bernabeo
Production Manager: Michael Esposito

MTM PUBLISHING
President: Valerie Tomaselli
Executive Editor: Hilary W. Poole
Editorial Coordinator: Tim Anderson
Editorial Assistants: Zachary Gajewski, Shalini Tripathi
Illustrator: Richard Garratt
Copyeditor: Carole Campbell

Design: Patrice Sheridan
Indexer: AEIOU, Inc.

Photographic Credits on page 1600.

VOLUME 5

JAMES T. RUSSELL

Inventor of the compact disc

1931–

James Russell developed the concepts underlying the compact disc (CD) and the digital video disc (DVD), both widely popular technologies that use light to store data. Russell's ideas were somewhat ahead of their time, and Russell has never profited materially from his inventions. Nonetheless, his work radically altered the computer, music, television, and film industries by making possible the digitization and optical storage of many different kinds of information.

EARLY YEARS

James Russell was born in Bremerton, Washington, in 1931. He became interested in science and engineering at a young age, and he was also a tremendous fan of classical music from childhood. He attended Reed College in Portland, Oregon, graduating with a degree in physics in 1953.

After graduating, Russell found a job with General Electric, which had contracted with the federal government to manage the Hanford nuclear plant, located in southern Washington. At Hanford, Russell became interested in creating new tools, designing, among other implements, a welder that used a beam of electrons.

SEARCHING FOR GOOD SOUND

The Hanford plant was built during World War II to provide plutonium and uranium for atomic and nuclear weapons. Because of its military importance and the sensitivity of the work, Hanford was located in a remote area of Washington state. Russell, who was living in nearby Richland, discovered that no radio stations in the area played classical music. To listen to the classical music he loved, Russell relied on his phonograph.

However, the phonograph was unsatisfactory on many levels; for one, records were played with heavy steel needles, which quickly wore out the records. Russell began exploring ways to make his records last

Russell was inspired to invent the CD because of his frustration with the needles used to play phonograph records.

longer, including using phonograph needles made of cactus spines. The spines did not ruin the records, and they produced a better sound than the steel needles, but cactus needles wore out quickly and had to be sharpened constantly.

Russell decided that the music industry needed a record-and-needle combination that would never wear out because the record and needle would never actually come into contact with each other. He thought the needle might be replaced with a laser.

DEVELOPING AN AUDIO RECORDER

Russell did not act on his idea until 1965, when Battelle Memorial Institute, a nonprofit research organization, took over management of Hanford from General Electric. Russell and the other staff members at Hanford became Battelle's employees, a change Russell anticipated would allow him to research his interest in lasers. He soon persuaded his new employers to let him develop his idea of creating an optical system that could be used to record and play music.

By 1970, Russell had created the first compact disc. As with today's CDs, the music was digitized into a code etched onto the disc using a laser. A laser was also used later to read the code, which was translated into musical sounds by a player. Russell's disc, however, differed in many ways from modern CDs; for example, the disc was about 12 inches (30 cm) in diameter.

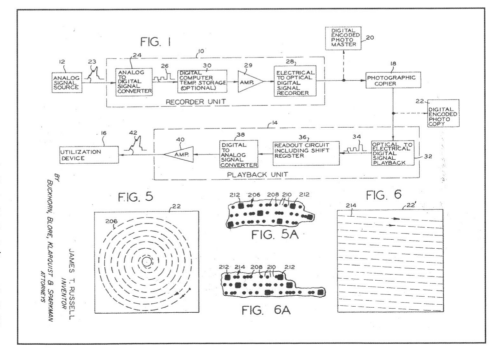

Illustrations from Russell's patent for an analog-to-digital recording and playback system; figure 1 is a diagram of the entire system; figures 5 and 6 are plan views of the digital encoding.

As Russell developed his audio system, he realized that if music could be digitized, so could video. Russell was very optimistic about the potential of this idea. "The vision I had in mind was of television programs on little plastic records," he later said. "The networks, instead of putting programs on television, would print records. And if you wanted to watch your favorite programs you'd get them in the mail and put in the disk whenever you want."

Music into Numbers

One of Russell's insights in developing the CD was the idea that something as flowing and amorphous as music could be represented digitally. He later recalled the reactions he received: " 'Music into numbers? Come on now, Russell.' When I first proposed it, it was not believed that you could digitize sound."

Previously, only analog recording had been used. In analog recording, a microphone picked up sound using a diaphragm that vibrated: the vibration was etched directly onto a record. When a record was played, a needle physically replicated that vibration, which was then translated back into sound. The problem was that the needle would eventually wear away the record's grooves. This led to a degradation in quality that prompted Russell to develop a system in which a laser "needle" would never actually touch the recording medium.

Such a system would rely on digital recording. The concept of recording music digitally struck Russell's contemporaries as unorthodox, but proved to be easy to accomplish. When a song or a sound is recorded on CD, the recorder samples the sound many times. In modern CDs, more than 44,000 samples are taken for each second of sound recorded.

Each sample is assigned a numeric value, which is then burned onto the CD with a laser. When the CD is played, a laser reads the numeric value of the many samples, and the CD player translates those values back into sound. Because so many samples are taken, the result is an extremely accurate sound recording that, if played properly, is never degraded by being scratched with a needle.

Much to Russell's frustration Battelle lost interest in developing his optical-storage ideas. A venture capitalist, Eli Jacobs, did show interest, however. Jacobs founded Digital Recording Corporation in 1971 to fund Russell's research into video discs. Russell created a working video recorder and player that was unveiled in 1974.

While working on the video recorder, Russell filed for a patent on a technique that is still used in compact discs. The technique allows a player to read data on discs in the correct order, a task called synchronization.

PATENTS

The video player Russell developed attracted a good deal of interest; representatives from several large electronics companies, including Sony Corporation and Royal Philips Electronics, were sent to evaluate the technology. At the time, companies like Philips had developed a laser optical disc that, unlike Russell's technology, which digitized audio and video to make it clearer, relied on analog recordings and was often of poor quality.

None of the companies that had been interested in Russell's technology proved willing to partner with Digital Recording, however, and

Shopping for CDs in Dhaka, Bangladesh, in 2003.

the firm began to founder. In 1985, Digital Recording went bankrupt, and its intellectual property, including Russell's patents, was acquired by a firm in Toronto, Canada, Optical Recording Corporation. The company offered the inventor a two-year contract as a consultant, which he accepted, moving to Toronto.

By then, compact discs were becoming a big business. Sony and Philips had introduced the first commercially available CDs in 1982. Although sales were initially slow, the improved sound quality of digitized music quickly gained the attention of audiophiles. In addition, compact discs began to interest the computer industry because they could store a great deal of information in a relatively small space.

Sony and Philips significantly modified Russell's original design: their CDs were much smaller and were in many ways more sophisticated. Nonetheless, the underlying technology was extremely similar to Russell's ideas, which had been protected by more than two dozen patents.

Russell's patents had been controlled by his previous employers, who had not gone to much trouble to enforce them, most likely because the compact disc was still a very new technology. As the compact disc became increasingly popular, however, Optical Recording decided to take far more aggressive action to protect Russell's patents.

In 1986, Optical Recording notified every other company in the compact disc industry of the patents it controlled, offering them the chance to negotiate a settlement in lieu of taking the case to court. Two years later, the company reached its first major licensing agreement with Sony; agreements with Philips and many other large companies followed.

One holdout was the entertainment giant Time Warner, which was manufacturing music CDs using Russell's synchronization technique. Optical Recording took Time Warner to court, and in June 1992 a federal jury found that Time Warner had violated Optical Recording's patent. Time Warner, which had by then manufactured 450 million compact discs, was ordered to pay Optical Recording $30 million.

> Although an inventor's life is often frustrating, particularly as one tries to convince others of the feasibility of an idea, it is also enormously exciting and in the end satisfying to create a useful device, a new technology, and even, on occasion, a new industry.
>
> —James Russell

OPTICAL MEMORY, CDs, AND DVDs

Russell, however, received none of that money. His two-year contract with Optical Recording had expired before the company reached its first licensing agreement, and he had moved back to Washington.

Russell worked as a consultant, establishing a company called Ioptics in Bellevue, Washington, in 1991. Ioptics focused on developing computer memory systems that used light; such systems would store

In the 1990s, CDs largely replaced cassettes and albums as the medium of choice among music buyers. However, a 198.5 percent increase in sales between 2004 and 2005 indicates that downloadable music files may eventually replace the CD.

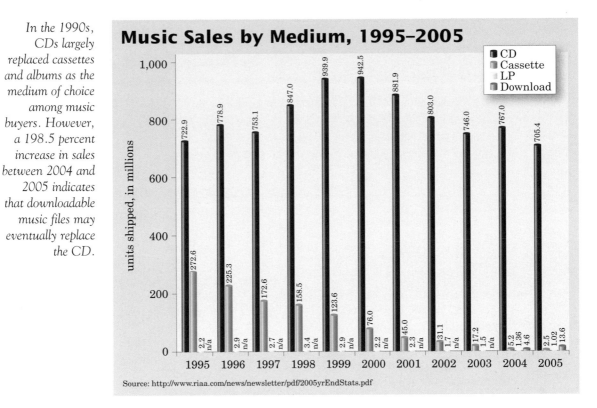

Music Sales by Medium, 1995–2005

Legend: CD, Cassette, LP, Download

units shipped, in millions

Year	CD	Cassette	LP	Download
1995	722.9	272.6	2.2	n/a
1996	778.9	225.3	2.9	n/a
1997	753.1	172.6	2.7	n/a
1998	847.0	158.5	3.4	n/a
1999	939.9	123.6	2.9	n/a
2000	942.5	76.0	2.2	n/a
2001	881.9	45.0	2.3	n/a
2002	803.0	31.1	1.7	n/a
2003	746.0	17.2	1.5	n/a
2004	767.0	5.2	1.36	4.6
2005	705.4	2.5	1.02	13.6

Source: http://www.riaa.com/news/newsletter/pdf/2005yrEndStats.pdf

more data and be faster than conventional memory storage. The low cost of traditional memory and production problems at Ioptics combined to drive the firm out of business in 1999. Russell continues to live in Bellevue, where he works as an optics consultant.

Russell's patents expired in the early 1990s, a few years before the introduction of the DVD made his idea of television programs on records a reality. The CD had largely replaced the vinyl record as the dominant format for music and the floppy disk as the dominant format for software. The DVD would do the same, largely replacing videotapes. By 2004, around 70 percent of U.S. households owned a DVD player and DVDs—in part because of services that, as Russell had anticipated, mailed discs to people's homes for a rental fee. These services continue to challenge the movie and television industries.

—Mary Sisson

Further Reading

Book

Oppenheim, Charles, ed. *Applications of Optical Media*. London: Aslib, 1993.

TIME LINE

1931	1953	1965	1971	1982	1991	1999
James Russell born in Bremerton, Washington.	Russell graduates from Reed College with a degree in physics.	Russell begins working on a digital recording and playback system.	Russell creates a video recorder.	CDs become commercially available.	Russell founds Ioptics.	Ioptics closes; Russell works as a consultant in the field of optics.

Web sites

The Discoverer

> An online article about Russell from *Reed* magazine.
> http://web.reed.edu/reed_magazine/Nov2000/a_the_discoverer/index.html

James T. Russell

> A profile of the inventor from the Lemelson-MIT Program.
> http://web.mit.edu/invent/iow/russell.html

See also: Communications; Edison, Thomas; Entertainment; Goldmark, Peter.

CLARENCE SAUNDERS

Inventor of the self-service
grocery store

1851–1953

Until the early 20th century, shopping at a grocery store was a far different experience from what it is now. Much of that difference is the result of the work of Clarence Saunders, an eccentric Memphis, Tennessee, businessman who patented the first self-service grocery store in 1917. His innovations have since become the hallmark of nearly every modern grocery store. Saunders not only changed the nature of grocery shopping but helped usher in the era of the self-service economy.

EARLY YEARS

Clarence Saunders was born in Virginia in 1851, but his family relocated to Tennessee when he was still young. His family was not wealthy, and at age 14, he left school to clerk in a general store.

By his twenties, Saunders was working for a wholesale grocer in Memphis, selling products directly to various grocers. Later he became a citywide salesman for another wholesaler. His experiences with grocers taught him that credit losses, high overhead, and inefficiency—common features of the early-20th-century grocery store—often doomed a small business to failure.

PIGGLY WIGGLY

On September 16, 1916, Saunders opened his own grocery store, Piggly Wiggly. There are various stories about the origins of the name, but Saunders's standard answer to the question, "Why did you name it Piggly Wiggly?" was, "So people will ask that very question!" Located at 79 Jefferson Street in downtown Memphis, it was unlike any store of its time.

Clarence Saunders is shown stocking shelves at his Keydoozle grocery store in Memphis, Tennessee, in 1937.

Typically, people shopped for groceries and other goods in a market or general store. Shoppers would enter the store and ask the clerk for the goods they wanted. The clerk stood behind a counter, along with the store's entire inventory. He would pull items from the shelves; measure out quantities from the jars, boxes, or barrels; then tally the costs and add the total to a customer's account. This process translated into lengthy interactions, high labor costs, and, in general, higher prices for customers.

Piggly Wiggly was different. It had shopping baskets, open shelves, checkout stands, and turnstiles (an invention patented by Saunders) at every entrance. Customers would walk along the aisles and select groceries themselves, without the aid of a clerk. Each item was marked with its price.

Saunders added other features that changed the shopping experience, including a uniformed staff, elaborate aisle displays, and many nationally advertised products. He instituted one-way aisles to smooth the flow of traffic and to force customers to view all the products. By cutting down on labor costs, he was able to sell goods more cheaply. He further appealed to cost-conscious shoppers with his cash-only policy. At other groceries, the charges were added to customers' accounts, but at Piggly Wiggly all customers paid in cash at the time of purchase. One advertisement read, "Your dollar at Piggly Wiggly will not help pay the BAD DEBTS of others."

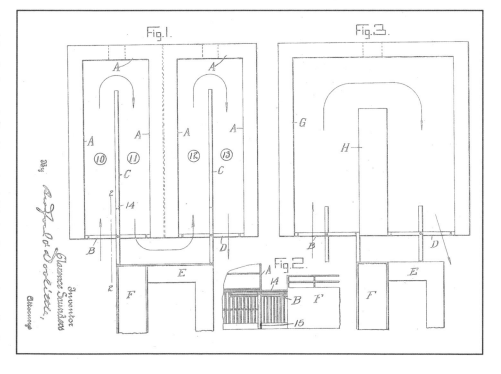

Illustrations from Saunders's 1917 patent for a self-service grocery store. Figure 1 shows an overhead view of several aisles of products; customers enter through a turnstile (B), and walk along a partition (C) past a series of cases (A) where products are displayed. Figure 3 shows a similar floor plan with fewer aisles.

Saunders submitted a six-page patent application for his innovations, and in 1917 received a patent for the "self-serving store," which he described as "distributing the merchandise of a store in such a manner that goods may be selected and taken by the customers themselves while making a circuitous path through the store."

Though many doubted Saunders at first, Piggly Wiggly quickly became a roaring success. Soon, new franchises opened up around the country, particularly in the South and Southeast. Each was identical—the same design and patented fixtures were used in every store. Saunders advertised his chain heavily, and the Piggly Wiggly name and logo grew to be not just recognizable, but beloved. By 1922, Piggly Wiggly Corporation was the second-largest grocery chain in the United States, with 1,200 stores in 29 states. The chain dominated the market in Kansas City, Missouri, and San Antonio, Texas. A decade later, Piggly Wiggly Corporation, now 2,600 stores strong, made more then $180 million per year.

THE DOWNFALL

Piggly Wiggly earned Saunders incredible wealth, and he was not afraid to display it. A well-known eccentric, Saunders commissioned a 38,000-square-foot (3,530-sq-m) mansion in Memphis to be built from pink marble. It was to feature an indoor pool and a shooting range. A dispute with stockbrokers in New York, however, brought an

A line of checkout clerks at a newly opened Piggly Wiggly in Encino, California, in 1962.

TIME LINE

1851	1916	1917	1922	1932	1937	1953
Clarence Saunders born in Virginia; his family soon relocates to Tennessee.	Saunders opens the first Piggly Wiggly store.	Saunders patents his self-service store plan.	Piggly Wiggly is the second-largest grocery chain in the United States.	Saunders is forced to declare bankruptcy.	Saunders founds the Keydoozle store.	Saunders dies.

end to Saunders's reign at Piggly Wiggly, and Saunders never slept a night in his dream mansion.

In 1932, stockbrokers tried to bring down the price of Piggly Wiggly stock, which had been on the market for some time. Incensed, Saunders allegedly took a train to New York with a suitcase full of cash, hoping to buy back the stock in his own company. In the end, Saunders was forced to declare bankruptcy and give up control of Piggly Wiggly Corporation. His ornate marble house was donated to the city and turned into the Pink Palace Museum.

STORES GALORE

Saunders went on to create other grocery store chains, with an eye toward automation. The next chain was called, oddly enough, "Clarence Saunders, Sole Owner of My Name Stores." (Saunders was reacting, it is believed, to his having to give up all rights to the name he had coined, Piggly Wiggly.) Although reasonably successful, the chain was forced to close during the Great Depression. Then, in 1937, Saunders founded another Memphis grocery, Keydoozle.

With Keydoozle, a play on "Key Does All," Saunders took the idea of self-service to another level. Keydoozle was one of the first mechanical, automated grocery stores—not unlike a very large vending machine. All the merchandise in the store was displayed in a series of glass cabinets. The customer indicated how many of each item she or he wanted by pushing a special key into a keyhole and pulling the trigger to indicate quantity. This information was recorded on a paper tape. Machinery in the back of the store was used to assemble the order and to send the products, via conveyor belt, to the checkout stand. The tape was used to tally the bill, and the groceries were boxed and ready to go. Keydoozle used space and human labor more efficiently, but the machinery, much of which Saunders built himself, was faulty and the store ultimately failed.

The notion of selecting one's own food at a supermarket is taken for granted in the 21st century, but it was invented by Saunders.

Undaunted even as he neared his 100th birthday, Saunders began drawing up plans for yet another store, the Foodelectric. This store held the promise of full automation. In a 1950s business book, *TNT: The Power within You*, Saunders described his plans for Foodelectric to the author: "The store operates so automatically that the customer can collect her groceries herself, wrap them and act as her own cashier. It eliminates the checkout crush, cuts overhead expenses and enables a small staff to handle a tremendous volume. . . . I can handle a $2 million volume with only eight employees." However, the store, which was to be located two blocks from the first Piggly Wiggly store in downtown Memphis, never opened. Clarence Saunders died in October 1953.

SELF-SERVICE SAGE

Saunders's vision of automation was clearly ahead of its time, as was his interest in and enthusiasm for self-service. Indeed, the first Piggly Wiggly predated other stores of its kind by at least two decades. At the time Saunders was dreaming of the Foodelectric, the supermarket, carrying all the hallmarks of Piggly Wiggly grocery stores, had just started to become a regular part of American life.

Saunders did not live to see the rise of the self-service economy—from the birth of laundromats, cafeterias, and self-service car washes to more recent high-tech examples, such as online shopping or airport kiosks that allow self-service check-in. Fifty years after Saunders's death, his vision has, in some ways, become reality. Fully automated checkout counters exist in most supermarkets; they are equipped with scanners,

Self-service Grocery versus Supermarket

Clarence Saunders is acknowledged as the inventor of the self-service grocery. Surprisingly, though, he is not the man behind the supermarket.

For many years, there was debate about the true founder of the supermarket. Then, in 1980, the Food Marketing Institute, in conjunction with the Smithsonian Institution and with funding from the H.J. Heinz Corporation (of Heinz ketchup fame), set out to settle the question. According to the Smithsonian's researchers, the first supermarket was invented by Michael Cullen, who opened a King Kullen store in Jamaica, Queens, New York, in 1930. Although Saunders had brought the world self-service, uniform stores, and nationwide marketing, Cullen was the one who built on that idea by adding separate food departments, selling large volumes of food at discount prices, and adding a parking lot.

credit card machines, and computer systems that talk to the customers. Meanwhile, his first grocery store, Piggly Wiggly, remains a leading chain in many parts of the United States.

—Laura Lambert

Further Reading

Articles

Carrns, Ann. "Piggly Wiggly Trims the Unsavory Fat off Its Aging Porker." *Wall Street Journal,* November 16, 1998.

Freeman, Mike. "Clarence Saunders: The Piggly Wiggly Man." *Tennessee Historical Quarterly* 51 (1992): 161–169.

Web sites

Piggly Wiggly Corporation
    Piggly Wiggly company history.
    http://www.pigglywiggly.com/
Society of Entrepreneurs: Clarence Saunders
    Short biography of Clarence Saunders.
    http://www.societyofentrepreneurs.com/hall_honor/saunders.asp

See also: Birdseye, Clarence; Food and Agriculture.

ADOLPHE SAX

Inventor of the saxophone

1814–1894

Belgian instrument maker Adolphe Sax achieved the rare feat of inventing an entirely new musical instrument around 1840 when he created the
saxophone. Sax also made improved versions of many existing wind
instruments. His entire body of work on musical instruments, rather than
the saxophone in particular, made him famous in his own lifetime.
Although Sax died before the word *jazz* was coined, his invention enjoyed
phenomenal success in the 20th century, largely attributable to its prominent presence in jazz performances.

EARLY YEARS

Adolphe Sax's future career owed much to his father Charles, who was a skilled woodworker before he turned his talents to making musical instruments. By 1820, Charles was running a thriving business in the Belgian city of Brussels. He and his wife had several children, including a son born on November 6, 1814, who, although christened Antoine-Joseph, was always known as Adolphe. Adolphe grew up working closely with his father in the family business.

In the early 1800s, musicians were showing great interest in improving the quality of wind instruments, which were primitive by today's standards. Military bands, for example, were often embarrassed by poorly designed and ill-performing instruments. At the time, composers, including Ludwig van Beethoven (1770–1827), were writing a new kind of classical music in which the sounds of individual instruments were increasingly important to the overall effect.

Although little is known about the depth of the collaboration between Adolphe and his father, they seem to have discussed ways to make better instruments and thus increase the success of their business. Recognizing his son's talent, Charles made sure that Adolphe received a good musical education. Adolphe Sax became a highly accomplished performer on a variety of wind instruments. His musical abilities were especially useful later, when he needed to demonstrate instruments to potential customers.

Wind instruments of the time were particularly poor in the bass (lower-pitch) range. Adolphe Sax's first major innovation, made in 1838, was a much-improved bass clarinet. This instrument was widely praised when he demonstrated it in Paris, a musical center in Europe at the time, during a visit in 1839. On this trip Adolphe made many useful contacts, among them the French composer Hector Berlioz (1803–1869), who became a lifelong friend and supporter.

A portrait of Adolphe Sax from the 1840s.

SUCCESS IN PARIS

Adolphe realized that his main career opportunities lay in the French capital, and so he reluctantly left his father's business to settle in Paris in 1842. He set up a factory to make wind instruments; the factory was soon in full production. His friend Berlioz wrote the first of several newspaper articles praising Sax's work. French generals and even the king of France started taking an enthusiastic interest in how Sax's creations might transform the quality of French military music.

Sax was much less popular with the existing musical instrument manufacturers of Paris, who resented his presence. Over the years his rivals challenged him in the court many times over his patents and other matters. Evidence exists of a plot to murder him.

Although Sax patented his saxophone in 1846, he had invented it several years earlier—possibly as early as 1838. Sax originally created the saxophone as a bass instrument, although he soon developed the smaller models that are most common today, such as the alto and tenor saxophones. Musical historians speculate that he came up with the concept of the saxophone through experiments in which he attached a clarinet mouthpiece to a brass instrument called an ophicleide. Like the saxophone, the ophicleide had holes down the side that the player could open and close with levers to produce different notes. Berlioz introduced the wider public to the saxophone in an 1842 article in which he praised its beautiful tone.

For use in military bands, the saxophone combined the advantages of the clarinet (ability to play fast melodies easily) with those of the trumpet and trombone (high decibel level). Sax concurrently developed a group of true brass instruments called saxhorns. He built these in various sizes, designing them to blend with one another musically. In a competition organized by the French government in 1846, Sax's instruments were judged to be the best, and his saxhorns and saxophones were adopted by all French military bands.

LATER YEARS

Sax's later life was not uncomplicated, although he continued to work hard and to come up with new ideas. Always a better inventor than businessman, he went bankrupt for the first time in 1852, though he later recovered from his losses. Sax suffered from lip cancer in the 1850s, then made a surprising recovery.

Adoption of the saxophone by the French government resulted in a need for instructors; Sax was appointed to an education post at the

Wind Instrument Basics

All wind instruments consist basically of an air-filled tube (either straight or curved) that the player blows into. Apart from the saxophone, wind instruments include oboes, clarinets, trumpets, and many others. The energy from the player's breathing is converted into sound waves that resonate (build up at a particular frequency or pitch) in the air of the tube. The longer the tube, the lower the notes that can be produced.

Wind instruments are divided into brass wind (such as the trumpet) and woodwind (such as the clarinet). In "true" brass instruments, the player's vibrating lips start and maintain the resonance within the tube. Woodwind instruments create resonance in other ways, usually by use of a reed, a thin sliver of material that vibrates with the player's breathing.

The saxophone, like the clarinet, has a mouthpiece with a single reed. Sax designed the inside shape of a saxophone's tube (the bore) to be conical—that is, it gets gradually wider between the mouthpiece and the bell (the open end), unlike a clarinet's bore, which is the same throughout its length. This difference in design is the main reason for the difference in sound between the two instruments. The saxophone's bore is also considerably wider overall, so that much more vibrating air is inside it. This helps the saxophone to be the loudest of all woodwind instruments, able to hold its own against trumpets and trombones.

As with a clarinet, different notes on a saxophone are achieved by covering and uncovering holes on the side of the instrument, using metal levers (keys). Uncovering a given hole has roughly the same effect as shortening the whole tube at that point; this is why a higher note is produced.

A saxophone maker in his workshop in 2005.

The saxophone continues to be a central instrument in jazz, particularly the New Orleans–based style called Dixieland.

music academy in Paris. He also commissioned works for the saxophone from classical composers. Sax continued to exhibit his newest instruments regularly at trade exhibitions. One of his most unusual, but least successful, inventions was a trombone that sprouted no fewer than 13 bells (the flared end of the instrument) instead of the usual one.

Life became more difficult in 1870 after Sax's patron, the French emperor Napoleon III, went into exile. Sax lost his job teaching the saxophone, and he went bankrupt a second time in 1873. His long old age was spent in relative poverty and obscurity, although his business survived under his son until the early 20th century.

TIME LINE

1814	1838	1842	1846	1852	1873	1894
Adophe Sax born in Brussels.	Sax creates an improved bass clarinet.	The saxophone wins the praise of composer Hector Berlioz.	Sax's instruments win a national contest in France.	Sax declares bankruptcy.	Sax declares bankruptcy for the second time.	Sax dies.

THE SAXOPHONE AFTER SAX

By the time Sax died in 1894, the saxophone was becoming increasingly important in military and other wind bands, including the famous American marching band led by John Philip Sousa (1854–1932). The sax was also commonly played in the variety theaters of the time. Early in the 20th century, jazz came on the scene. The first New Orleans jazz bands did not use saxophones; in fact, the jazz world initially adopted them reluctantly, under pressure from concert promoters. However, in the 1920s a "saxophone craze" swept the United States. New styles of jazz emerged that better suited the saxophone, both as a solo instrument and in larger groupings. The greatest jazz saxophonists include Charlie Parker, Lester Young, Ornette Coleman, Sonny Rollins, and John Coltrane, to name just a few; important contemporary saxophonists include David Sanborn and Branford Marsalis. The saxophone continues to be a popular instrument and has remained basically as Adophe Sax created it.

—Richard Beatty

Further Reading

Horwood, Wally. *Adolphe Sax, 1814–1894: His Life and Legacy.* Hertfordshire, UK: Egon, 1992.

Ingham, Richard, ed. *The Cambridge Companion to the Saxophone.* Cambridge: Cambridge University Press, 1999.

See also: Cristofori, Bartolomeo; Entertainment.

ARTHUR SCHAWLOW AND CHARLES TOWNES

Inventors of laser technology
1921–1999 and 1915–

In 1958, Arthur Schawlow and Charles Townes published an exposition of how to make lasers. Their article enabled the creation of a technology that has helped advance a wide variety of fields, including manufacturing, surveying, astronomy, and medicine. Both men received Nobel Prizes for their contributions.

EARLY YEARS: TOWNES

Charles Townes was born in Greenville, South Carolina, in 1915. Townes's father was a lawyer who also owned a farm, and Townes later recalled being fascinated as a child by the natural world and by the machinery on the farm.

Townes was a fast learner, skipping grades in school and entering Furman University in Greenville at the age of 16. Furman was not well known, but attending the university allowed Townes to live at home, an arrangement that helped his family save money, an important consideration in 1931, during the Great Depression.

Townes enjoyed Furman, discovering physics in his sophomore year. He graduated in 1935 with a bachelor of arts degree in modern languages and a bachelor of science degree in physics.

THE QUEST FOR FELLOWSHIPS

Townes wanted to study physics in graduate school, but his family could not afford to pay his tuition. He applied for fellowships, receiving one at Duke University in Durham, North Carolina, where he finished a master's degree in 1937. He did not, however, get a fellowship allowing him to study for a doctorate.

Townes applied to other schools for fellowships and was turned down. At length, he decided to get into the best program possible and to worry about the money later. He applied and was accepted at the California Institute of Technology. Eventually Townes was appointed to a teaching assistantship that helped pay for his education. He was awarded his doctorate in physics in 1939.

Townes wanted to find a position in academia, but during the Great Depression teaching jobs were virtually impossible to find. Instead, a professor arranged for Townes to work for Bell Telephone Laboratories in New York City. During 1939 and 1940, while

Charles Townes in 1961.

Townes was working for Bell on a variety of projects, World War II broke out, and Nazi Germany conquered much of western Europe. The United States did not enter the war until December 1941, but the U.S. government realized that a war with Germany was probably inevitable and began working with various corporations, including Bell, to improve defense technology.

RADAR AND MICROWAVES

Townes went to work on radar systems, which sparked his interest in microwave spectroscopy—the use of microwaves to study molecules. After World War II ended in 1945, Townes persuaded Bell to allow him to work on microwave spectroscopy, but Townes's work was soon seen to be more academic than industrial.

Microwave Spectroscopy

In early 1941, while Townes was working for Bell, he was assigned to help design bombing systems that would use radar to locate targets for the U.S. military. At the time, radar systems were fairly primitive; the need to improve the technology to locate targets with more precision was obvious.

Radar systems work by bouncing radio waves off a target, then analyzing the results. One strategy to improve radar that Townes developed was creating radar systems that used shorter radio waves, which were supposed to provide more detailed pictures.

Townes worked on systems using shorter and shorter waves, finally being assigned to a system that used a wave only a half inch (1.25 cm) in length. While working with such waves, Townes realized that they would be absorbed by water vapor in the air, which would render the radar system useless because the radio waves could not bounce back to the receiver. Despite his objections, the system was built and failed in the field.

Townes was nonetheless intrigued by the molecule's ability to absorb waves of certain lengths, an effect especially pronounced in the short radio waves called microwaves. Townes realized that this phenomenon could be developed into a method to study the structure and properties of molecules; this insight would eventually lead to the entire field of microwave spectroscopy.

In 1948, Townes accepted the position of associate professor in the physics department of Columbia University, also in New York City. A year later, he hired Arthur Schawlow, a young postdoctoral fellow from Toronto, to assist him.

EARLY YEARS: SCHAWLOW

Like Townes, Arthur Schawlow was a youthful prodigy. He was born in Mount Vernon, New York, in 1921, but his family moved to Toronto, Canada, three years later. Schawlow's mother, who had suggested the move, was from Canada; his father, an insurance agent, was an immigrant from what is now Latvia.

Schawlow did very well as a student and entered a school for gifted children. When he graduated from high school in 1937 his family could not afford to pay college tuition. Schawlow was able to win a scholarship to the University of Toronto that allowed him to study mathematics or physics. He studied physics and decided to remain in the field.

Schawlow graduated from the university in 1941, after Canada had entered World War II. Any further studies were suspended during the war, so Schawlow taught classes to military personnel and worked on microwave antennas for a radar factory.

When the war ended in 1945, Schawlow went back to the University of Toronto to pursue his doctorate. Between the war and the Great Depression, the physics department had been seriously depleted of personnel and equipment, but the optics program was still operating, so Schawlow studied optical spectroscopy. He received his doctorate in 1949 and was awarded a fellowship at Columbia under Townes.

THE MASER

The two men got along well, working together on microwave spectroscopy and writing a textbook on the subject published in 1955. Townes was working on

Arthur Schawlow with an "eraser laser" in 1968.

Townes and
Gordon present
the maser at
Columbia
University in
1955.

other projects, one of which troubled him greatly. He wanted to create a tool that would generate very short microwaves—a millimeter in length—that could then be used to study molecules, but he could not find a way to do so.

In April 1951, Townes had a breakthrough. He realized he could make molecules emit microwaves. Molecules could absorb energy until they reached an excited state in which they could not absorb any more. Excited molecules could release their excess energy and return to their normal state.

Townes realized he could excite certain molecules and bombard them with even more energy. This step would force the excited molecules to give up their excess energy, a process called stimulated emission. When stimulated emission occurred, the molecules released more energy than they had absorbed. Although the energy used to bombard the molecules could be of any wavelength, the energy released would be of a specific wavelength because of the way molecules stored energy. Townes thought he could find molecules that would emit microwaves.

Townes shared the idea with Schawlow, but he did not pursue it. Schawlow married Townes's younger sister in 1951 and had to leave Columbia because the university did not want relatives working together. Schawlow went to work for Bell, and Townes put his Columbia students to work on his idea.

In the spring of 1954, Townes and a student, Jim Gordon, constructed the first device that could emit short microwaves. The device emitted microwaves that were very pure; that is, they were all of the same wavelength. After the microwaves were created through stimulated emission, they were directed into a resonance chamber where the waves were reflected back and forth. As the waves bounced around the chamber, they picked up even more strength. Since the chamber was a certain length, waves that were not the desired length dissipated, leaving those of the desired length to be amplified. Consequently, the device produced microwaves of all the same wavelength. Townes and Gordon named their device the maser (microwave amplification by stimulated emission of radiation).

THE LASER

The maser caused a sensation in scientific circles. Townes became interested in making what became known as a laser, a maser-like device that would amplify visible light rather than microwaves. Light waves were of much higher energy than radio waves, so lasers had more potential uses than masers. An intense, pure beam of light could, for example, vaporize dark material while leaving light material undamaged.

A problem arose: stimulated emission, although crucial to the maser process, was not the only process involved in the device's workings. The maser's resonance chamber also was crucial. Townes realized that any laser would also need a resonance chamber. Light waves, however, were much shorter than radio waves and would require a different type of resonance chamber. Townes had no success in designing a chamber for the proposed laser. In 1957 he presented the problem to his brother-in-law, who had a background in optics. Schawlow quickly determined how to build a resonance chamber for a laser.

Masers were a hot topic in the scientific world, so the two men decided to publish their ideas before trying to build a working laser. Their article outlining how to build a laser appeared in the journal *Physical Review* in December 1958. The article sparked a race among scientists to build the first laser; in May 1960, Theodore Maiman, a scientist at the technology company Hughes Laboratory, came in first. By the end of the year, at least four more laboratories had built lasers using Townes's and Schawlow's ideas.

LATER LIVES

By that point, Townes was no longer working on lasers. In 1959 he had become director of research at the Institute for Defense Analysis, an organization in Washington, D.C., that advised the U.S. military on science and technology. Two years later, he was appointed professor of physics and

The Wave Spectrum

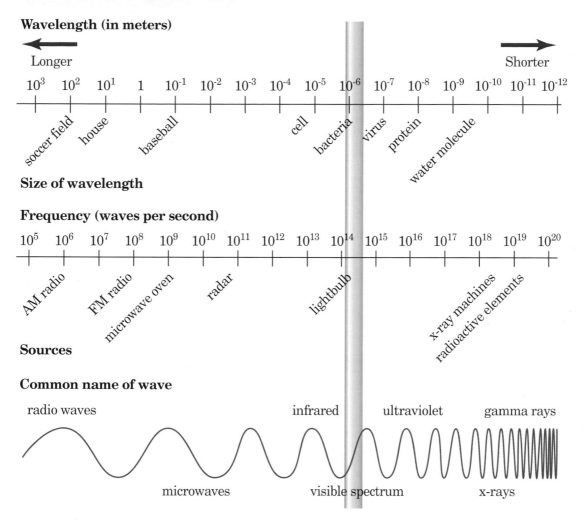

Wavelength (in meters)

Longer ← → Shorter

10^3 10^2 10^1 1 10^{-1} 10^{-2} 10^{-3} 10^{-4} 10^{-5} 10^{-6} 10^{-7} 10^{-8} 10^{-9} 10^{-10} 10^{-11} 10^{-12}

soccer field · house · baseball · cell · bacteria · virus · protein · water molecule

Size of wavelength

Frequency (waves per second)

10^5 10^6 10^7 10^8 10^9 10^{10} 10^{11} 10^{12} 10^{13} 10^{14} 10^{15} 10^{16} 10^{17} 10^{18} 10^{19} 10^{20}

AM radio · FM radio · microwave oven · radar · lightbulb · x-ray machines · radioactive elements

Sources

Common name of wave

radio waves · infrared · ultraviolet · gamma rays

microwaves · visible spectrum · x-rays

The wave spectrum is extremely vast, with differing wavelengths possessing different properties. The maser is derived from the microwave spectrum, whereas the laser is derived from the visible spectrum.

provost at the Massachusetts Institute of Technology in Cambridge, Massachusetts. In 1964, he won the Nobel Prize in Physics for his work developing the maser and the laser. Three years later, he moved to the University of California–Berkeley, where he became deeply involved in the field of astrophysics.

Schawlow continued to work on the laser at Bell; in 1961, he accepted a professorship at Stanford University in Stanford, California, where he headed the physics department from 1966 to 1970. He also helped develop the field of laser spectroscopy—the use of lasers to study molecules and atoms—for which he was rewarded the Nobel Prize in Physics in 1981. Schawlow remained at Stanford until his death in April 1999.

The Gould Affair

Since Schawlow and Townes were working for Bell when they originated the design for the laser, Bell filed to patent the device in July 1958. In April of the next year, however, another patent for the laser was filed by Gordon Gould. Gould's patent application was denied, but he did not give up, challenging the Bell patent for years on the ground that he had had the ideas for the laser first.

Gould was a doctoral student in physics at Columbia University in the 1950s. His notebooks suggest that he did realize some concepts behind the laser independently. Matters were complicated, however, by the fact that Gould and Townes had at least two conversations related to the development of the laser in late 1957, and each man accused the other of stealing ideas.

The Gould case was finally resolved in 1976, shortly after the Bell patent expired. Gould was granted a patent for part of the inner workings of the laser, which required companies using lasers to continue paying royalties on the technology until the Gould patent expired in the 1990s.

By the time of Schawlow's death, lasers were an estimated $3.5 billion industry. Whereas masers have limited uses, usually to amplify microwaves in radio astronomy, lasers are used in an enormous range of industries and devices. Lasers are used as precision cutting tools in com-

TIME LINE

1915	1921	1939	1948	1949	1954
Charles Townes born in Greenville, South Carolina.	Arthur Schawlow born in Mount Vernon, New York.	Townes earns his doctorate in physics from the California Institute of Technology.	Townes becomes an associate professor of physics at Columbia University.	Schawlow earns his doctorate in optics and is awarded a fellowship to Columbia.	Townes and Jim Gordon build a maser.

A laser blasts a pinpoint hole through a razor blade in a 1963 test.

puter manufacturing and in eye surgery. Laser scanners are used to read discs containing music, computer programs, and movies; they also are used to read the bar codes found on products in stores. They are used to measure distances in ordinary surveying and in astronomy. Lasers have gone from Townes's and Schawlow's conception to a device that enriches many aspects of modern life on a daily basis.

—Mary Sisson

TIME LINE

1958	1960	1964	1981	1999
Schawlow and Townes's article on lasers is published in *Physical Review*.	The first laser is built by Theodore Maiman.	Townes wins the Nobel Prize in Physics.	Schawlow wins the Nobel Prize in Physics.	Schawlow dies.

Further Reading

Books

Bromberg, Joan Lisa. *The Laser in America, 1950–1970*. Cambridge, MA: MIT Press, 1991.

Taylor, Nick. *Laser: The Inventor, the Nobel Laureate, and the Thirty-Year Patent War*. New York: Simon & Schuster, 2000.

Townes, Charles H. *How the Laser Happened: Adventures of a Scientist*. New York: Oxford University Press, 1999.

———. *Making Waves*. Woodbury, NY: AIP, 1995.

Web sites

Adventures of a Scientist: Conversation with Charles H. Townes, Nobel Laureate
    An interview with Townes from UC Berkeley's Institute of International Studies.
    http://globetrotter.berkeley.edu/people/Townes/

The Nobel Prize in Physics 1964
    Information on Townes by the Nobel Foundation.
    http://nobelprize.org/nobel_prizes/physics/laureates/1964/

The Nobel Prize in Physics 1981
    Information on Schawlow by the Nobel Foundation.
    http://nobelprize.org/nobel_prizes/physics/laureates/1981/

See also: Optics and Vision; Science, Technology, and Mathematics.

SCIENCE, TECHNOLOGY, AND MATHEMATICS

Science, technology, and mathematics are closely related. Science is a way of systematically figuring out how the world works. Mathematics, one of a scientist's most useful tools, is a method of using numbers and quantities. Technology is a way of applying science and mathematics to solve problems encountered by humans and to make work easier, safer, and more efficient.

Early humankind had no science or mathematics; its technology was primitive. All invention happened by accidental discovery or by trial and error. As aeons passed, science, mathematics, and technology were developed and have become increasingly important to invention. In the modern age, people speak about high technology, including lasers, fiber optics, and computers. Such inventions are based on detailed scientific and mathematical knowledge that have been built up over thousands of years.

THE TIME BEFORE SCIENCE

Humans have inhabited the earth for about five million years, but information has been recorded only for the last few thousand years. The vast majority of human experience is unrecorded. In prehistory—the time before the invention of writing—there were of course no books, newspapers, or other written materials in which inventions or discoveries could be recorded. During those millennia, every discovery someone made was potentially a useful invention, but no sure way was available to pass

A letter written in cuneiform about 2,300 years ago, found in Jerusalem.

useful knowledge on to people outside an immediate group except orally. Such technology as existed was developed by people making accidental discoveries about controlling their environment.

This process of sharing information began to change with the invention of writing. Written words were not invented until around 3500 BCE, when the Babylonians, from a region now part of Iraq, started making marks on clay tablets, in a system known as cuneiform, to record the words of the language they spoke. About a thousand years later, the ancient Egyptians invented the base 10 number system that is still used today, with symbols (hieroglyphs) representing 1, 10, 100, and so on. They used a picture of a tadpole to represent the number 100,000 (historians suggest that this may have been because tadpoles occur in large numbers) and an image of a seated god to represent 1,000,000, a large number almost infinite to them and a sign of the god's great power. Written words and numbers were a way to record knowledge and experience and made mathematics and science possible.

THE BEGINNINGS OF SCIENCE

The ancient Greeks are sometimes described as the first real scientists: they tried to understand the world around them by thinking and reasoning. Greek philosophers (literally "wisdom lovers") were pioneers of mathematics and used it to lay the foundations of western science. Unlike modern scientists, however, they did not carry out systematic experiments to test their theories. Most of their scientific ideas evolved from mathematics. Thus, the Greek mathematician Pythagoras (ca. 582–507 BCE) believed "number is the within of all things" and taught how mathematics could be used to understand the world. He was one of the fathers of geometry, and his students later used his mathematical ideas to reason that the earth was spherical and moved in a circular orbit around the sun. This conclusion was an early example of mathematics being used to advance science.

Greek philosopher and mathematician Archimedes (ca. 287–212 BCE) was known for developing some of the earliest scientific ideas about how objects move. These ideas were the basis for the science of mechanics, which is now a

Engraving from 1833 depicts Pythagoras, with stick, explaining his theorem of right-angled triangles.

A contemporary student solves a problem with algebra, pioneered by al-Khwarizmi.

part of physics. Through studying water, Archimedes invented the science of hydrostatics, which explains buoyancy—the ability of materials less dense than water to float. His various inventions included an auger device for lifting water called Archimedes screw, and new types of levers and pulleys—all machines that use mathematics and science in the service of technology. Archimedes reputedly cared little for practical inventions; he was more interested in thoughts and ideas.

Greek theories were passed on to the Romans by conquest and trade; the Roman Empire lasted from around 29 BCE to 495 CE. Although the Romans were less interested in science than the Greeks and made fewer scientific discoveries, they turned many of the Greeks' ideas into practical technologies. Although the Greeks invented gears (wheels with teeth cut into their edges that could increase the speed or power of a machine), the Romans were the first to use them extensively.

THE MIDDLE AGES: SCIENCE WITHOUT SYSTEM

After the Roman Empire fell to invaders from the north and east, Europe entered a period sometimes referred to as the Dark Ages. Much of the order that the Romans had introduced and imposed disappeared, society became more chaotic, and many scientific ideas were lost.

Mathematics and science continued to develop in the Islamic world, which was then expanding from its traditional base in the Middle East to incorporate a vast area that stretched from Spain to China. Islamic scholars gradually acquired scientific knowledge from Europe, India, and China and translated it into their own languages. During the Middle Ages (which lasted from the fall of the Roman Empire to about the fifteenth century), Islamic mathematicians built on the knowledge they had gathered. Islamic mathematician Muhammad ibn Musa al-Khwarizmi (ca. 780–850 CE) combined ancient Greek and Indian ideas to develop algebra, which gets its name from the Arabic term *al-jabr*, meaning "completion." Islamic knowledge gradually filtered back to Europe from the twelfth century on, and the ideas that had been lost during the Dark Ages were restored in a much-improved form.

Although scientific and mathematical ideas were spreading throughout the world, they were far short of the modern body of science that people use today. Many scientific ideas were gradually coming together from different sources but had not been organized into a complete system.

Some of the most important inventions from this period developed with little or no help from science and math. For example, improved horse harnesses for plows made working the fields easier and greatly increased agricultural productivity. Gunpowder was invented in China, where the mariner's compass, paper, and basic printing technology also were invented. In the hands of German printer Johannes Gutenberg (ca. 1400–1468), printing technologies developed into the modern printing press in the 1450s. Gutenberg's invention spawned an enormous increase in books, libraries, and education, enabling knowledge to spread from a few religious scholars to a wider population.

THE SCIENTIFIC REVOLUTION

In modern times, scientific discoveries produce new technologies, but hundreds of years ago science and technology often worked the opposite way. A new invention often helped push science forward: for example, the invention of telescopes in the seventeenth century.

The first telescope was made by Dutch lensmaker Hans Lippershey (ca. 1570–ca. 1619) in 1608. Italian physicist Galileo Galilei (1564–1642) heard about the telescope soon after its invention and proceeded to make a much better telescope. Looking through it, Galileo made remarkable discoveries. What he saw in the sky quickly persuaded him that earth was not the center of the universe, as most people in his time believed. Instead, he realized that the sun was at the center of our solar system and that the earth and the other planets revolved around it. This idea, called the heliocentric (sun-centered) theory, had been put forward in 1543 by Nicolaus Copernicus (1473–1543). When Galileo started to promote the heliocentric theory, he came into conflict with the Roman Catholic church; Galileo was sentenced to house arrest in 1633. Nevertheless, Galileo's battles with the church authorities marked the beginnings of a revolution: new scientific theories were starting to take the place of long-held religious beliefs.

Others joined Galileo in putting the human world on a more scientific footing. Dutch mathematician Christian Huygens (1629–1695) made better telescopes and further astronomical discoveries. In 1656, he also invented the pendulum clock, a crucial piece of technology that made accurate timekeeping possible. This major step forward allowed scientists to measure the rate at which things happened. Another great scientist of this period, Englishman Robert Hooke (1635–1703), invented the compound microscope (one with several lenses) around 1665 and dis-

The Galileo spacecraft captured images of a large storm brewing in Jupiter's atmosphere in February 1998.

covered the microscopic world of living organisms: he invented the word *cell* to describe what he saw.

The telescope, the microscope, and the pendulum clock were technologies that helped others to drive science forward. One scientist who rode this wave of progress was Englishman Isaac Newton (1642–1727). Building on Galileo's work, Newton used mirrors to make a more powerful kind of telescope (a reflector telescope). This invention helped him devise the first comprehensive theory of gravity (a force of attraction that pulls objects in the universe together). Because of this theory and his work on light, Newton is considered by many to be the greatest scientist of all time.

Newton was also a pioneer of calculus, an advanced form of math that was simultaneously developed in Germany by mathematician and philosopher Gottfried Wilhem Leibniz (1646–1716). Calculus, which is used to measure rates of change, has been an essential tool for scientists since its invention. Among his many other achievements was a calculator: in the 1670s Leibniz was one of the first to build a mechanical calculating machine, a technology that helped to overcome the limitations of the human mind. Later mathematicians, including Englishman Charles Babbage (1792–1871), extended this idea greatly, continuing a slow process of invention that gradually resulted in the modern computer.

Mathematical Inventions

Technology generally makes life easier. That is certainly true of the inventions people have devised to help them with math. One of the earliest mathematical inventions, the abacus, is a simple calculator that dates to around 500 BCE. By the seventeenth century, mathematicians Gottfied Leibniz and Frenchman Blaise Pascal (1623–1662) were making some of the first mechanical calculators using simple arrangements of gears. In the 18th century, English mathematician Charles Babbage tried to do for mathematics what the steam engine had done for factory work and transportation: he wanted to build enormous engines that could carry out complex calculations. However, the hugely ambitious machines he designed needed thousands of handmade gears and were far too complex and expensive to be built in his lifetime.

Calculators became a dead end for mathematical technology for a time and other, older technologies prospered instead. The most important of these was the logarithm table, one of the greatest mathematical inventions of the seventeenth century, developed in 1614 by a Scotsman, John Napier (1550–1617). Clearly very pleased with his invention, Napier called it "a description of the admirable table of logarithms, with a declaration of the most plentiful, easy, and speedy use thereof in all kinds of trigonometry, as also in all mathematical calculations." Logarithms (also called logs) were a quick way of multiplying two or more numbers by adding smaller numbers, which could be looked up in a printed book (or log table). In 1617, Napier invented another mathematical technology—Napier's bones. It consisted of a set of short wooden rods on which numbers were printed. These could be placed next to one another to carry out complex multiplications.

Logarithms were also used by the slide rule, a simple calculator invented in 1654. This device was a wooden ruler, with a sliding middle section on which logarithms were printed. Calculations could be made by sliding the middle section back and forth. Slide rules were so simple and effective that they were widely used for more than three hundred years—even in schools—until electronic calculators made them obsolete in the 1970s.

THE BEGINNINGS OF MODERN SCIENCE AND TECHNOLOGY

Further inventions opened new avenues of scientific research. After German physicist Daniel Fahrenheit (1686–1736) invented the mercury thermometer in 1714, developing scientific theories of heat became possible. Mathematics was also crucially important in such work.

By this time, many people were studying science. In the United States in the mid-18th century, Benjamin Franklin (1706–1790) retired from his profitable printing business to study what was then an intriguing novelty: electricity. His scientific research, which included a dangerous experiment flying a kite in a thunderstorm in 1752, led to a very important practical invention: the lightning rod. His research also helped others to see that electricity could be compared to a kind of fluid that could drift, like water, from one place to another. In Franklin's time, scientific knowledge could easily be published and passed from person to person or from country to country. Thus, following Franklin's work, Italian physicist Alessandro Volta (1745–1827) realized that moving electricity made a current. This understanding enabled him to invent the world's first battery in 1800.

Volta's work was an inspiration for English chemist Michael Faraday (1791–1867), who explored the connections between electricity and magnetism. His investigations led to the invention of electromagnets and electric motors, pioneered by another Englishman, William Sturgeon (1783–1850), in 1832. By the end of the 1800s, using this technology, Thomas Edison (1847–1931) had opened the world's first electric power plants. In little over a century, electricity had been transformed from a curiosity into a science, then into a practical technology that would change the lives of millions.

Physical sciences were not the only areas of progress: biology and chemistry also made great leaps forward during the 18th and 19th centuries. In 1735, Swedish botanist Carl von Linné (or Carolus Linnaeus, 1707–1778) developed a system for grouping plants and animals that were similar to one another, using words such as *kingdom* and *species*. These classification systems, which are still used today, greatly advanced research in botany (plant science) and zoology (animal science). Another important biological discovery was made in the mid-1800s by Austrian monk Gregor Mendel (1822–1884). His observations of plants through generations led him to the idea that the characteristics of

TIME LINE

3500 BCE	ca. 582 BCE	ca. 287 BCE	ca. 780 CE	1450s	1543
Writing invented by the Babylonians.	Pythagoras born.	Archimedes born.	Muhammad ibn Musa al-Khwarizmi born.	Johannes Gutenberg invents the printing press.	Nicolaus Copernicus advances the heliocentric theory.

The Periodic Table

group

	1	2	3	4	5	6	7	8	9	10	11	12	13	14	15	16	17	18
1	1 H																	2 He
2	3 Li	4 Be											5 B	6 C	7 N	8 O	9 F	10 Ne
3	11 Na	12 Mg											13 Al	14 Si	15 P	16 S	17 Cl	18 Ar
4	19 K	20 Ca	21 Sc	22 Ti	23 V	24 Cr	25 Mn	26 Fe	27 Co	28 Ni	29 Cu	30 Zn	31 Ga	32 Ge	33 As	34 Se	35 Br	36 Kr
5	37 Rb	38 Sr	39 Y	40 Zr	41 Nb	42 Mo	43 Tc	44 Ru	45 Rh	46 Pd	47 Ag	48 Cd	49 In	50 Sn	51 Sb	52 Te	53 I	54 Xe
6	55 Cs	56 Ba	* 71 Lu	72 Hf	73 Ta	74 W	75 Re	76 Os	77 Ir	78 Pt	79 Au	80 Hg	81 Tl	82 Pb	83 Bi	84 Po	85 At	86 Rn
7	87 Fr	88 Ra	† 103 Lr	104 Rf	105 Db	106 Sg	107 Bh	108 Hs	109 Mt	110 Ds	111 Rg	112 Uub	113 Uut	114 Uuq	115 Uup	116 Uuh	117 Uus	118 Uuo

period

		3	4	5	6	7	8	9	10	11	12	13	14	15	16	17
*Lanthanoids	*	57 La	58 Ce	59 Pr	60 Nd	61 Pm	62 Sm	63 Eu	64 Gd	65 Tb	66 Dy	67 Ho	68 Er	69 Tm	70 Yb
†Actinoids	†	89 Ac	90 Th	91 Pa	92 U	93 Np	94 Pu	95 Am	96 Cm	97 Bk	98 Cf	99 Es	100 Fm	101 Md	102 No

With the help of mathematics, Dmitry Mendeleyev constructed the periodic table of elements. Chemical elements in the table are organized into vertical columns (groups) and horizontal rows (periods), which have similar properties.

plants could be passed to the members of each generation; thus he pioneered the modern science of genetics. Similar advances would also play an important part in medicine.

Modern chemistry began in 1789 when French scientist Antoine Lavoisier (1743–1794) published a book about the chemical elements. About eighty years later, in 1868, Russian scientist Dmitry Mendeleyev (1834–1907) developed a classification system for all known chemical elements. Mendeleyev published *The Principles of Chemistry*, in which he showed such elements in an ordered grouping, the periodic table of the elements. This brilliant piece of science not only explained all the facts then known

TIME LINE (continued)

1608	1614	1633	1656	ca. 1665	1670s
Hans Lippershey builds a telescope.	John Napier invents the logarithm table.	Galileo Galilei sentenced to house arrest for his scientific ideas.	Christian Huygens designs a pendulum clock.	Robert Hooke invents the compound microscope.	Gottfried Leibniz invents a calculating machine.

about the chemical elements, but also highlighted gaps in knowledge that later research was able to fill.

Despite such great advances, scientific knowledge was still often divorced from practical technology and invention. Thus, when French artist Louis Daguerre (1789–1851) developed photography in the mid-19th century, he did so entirely by trial and error. Only later would the sciences of chemistry and physics be able to explain Daguerre's processes and accomplishments. Photography later became an essential tool for helping scientists to record their discoveries.

MODERN TIMES

During the 20th century, science and mathematics advanced to such a degree that they were able to inspire and bring to fruition previously unimaginable technologies. The best illustration is the development of one of the 20th century's greatest inventions: atomic energy. The first person to realize the possibility of making energy from atoms was Albert Einstein (1879–1955). Einstein was a theoretical physicist: instead of carrying out experiments, he used imaginary thought experiments and pages of complex math, including the calculus developed by Leibniz and Newton, to work on virtually all his scientific ideas.

Einstein's theories underpinned the work of Italian physicist Enrico Fermi (1901–1954), who proved in 1942 that splitting the nucleus (center) of an atom could make energy. Fermi's scientific work led immediately to nuclear technology, giving birth to a whole range of new technologies, including atomic bombs and nuclear power plants, smoke detectors, medical inventions such as NMR (nuclear magnetic resonance) scanners, and radiotherapy treatments for illnesses like cancer. The mathematics of Einstein's theories helped to develop the nuclear science of Fermi's research, which led to new technology.

Scientists see their work as exploration—none more so than American physicist Robert H. Goddard (1882–1945), father of the modern space rocket. Goddard's early attempts to send rockets above the earth's atmosphere (and thus outside the overwhelming pull of gravity) involved rigorous science and math; nevertheless, many people thought Goddard was crazy when he calmly proposed sending a rocket to the moon. Scientific theories sometimes counted for little until practical technologies proved them beyond all doubt. Not until the 1950s and 1960s, when orbiting

TIME LINE

1714	1735	1752	1789	1800	1800s
Daniel Fahrenheit invents the mercury thermometer.	Carl von Linné develops a classification system for plants and animals.	Benjamin Franklin conducts his famous kite experiment.	Antoine Lavoisier publishes a book about chemical elements.	Alessandro Volta invents the battery.	Gregor Mendel observes the inherited characteristics of plants and animals.

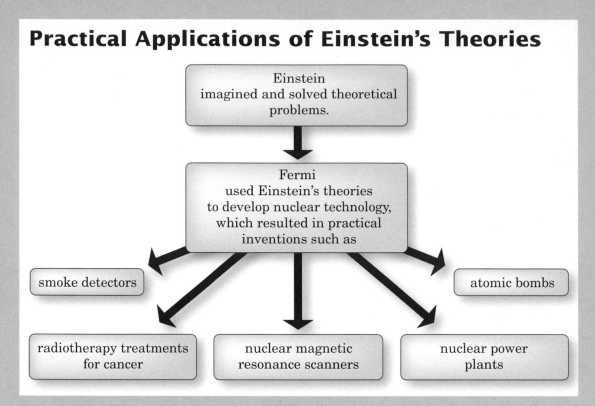

Practical Applications of Einstein's Theories

Einstein imagined and solved theoretical problems.

Fermi used Einstein's theories to develop nuclear technology, which resulted in practical inventions such as

- smoke detectors
- radiotherapy treatments for cancer
- nuclear magnetic resonance scanners
- nuclear power plants
- atomic bombs

The highly theoretical work of Albert Einstein was subsequently translated into practical uses by other scientists and inventors.

satellites and manned space expeditions were a reality, was Goddard's genius fully recognized.

Because of visionaries like Goddard, we now have an enormous and ever-growing amount of scientific knowledge about space. Indeed, scientists often say that more is known about the moon than about certain parts of the earth, particu-larly its deep oceans. Once widely used as a dumping ground for sewage and other waste, oceans are now a matter of grave environmental concern, with attitudes toward oceans having changed dramati-cally since the 1970s. French explorer Jacques-Yves Cousteau (1910–1997) and his partner Émile Gagnan played a cen-tral part in this change by inventing the

TIME LINE (continued)

1820s–1871	1832	1868	1942	1943
Charles Babbage describes how a programmable computer could work.	William Sturgeon invents an electric motor.	Dmitry Mendeleyev develops the periodic table of elements.	Enrico Fermi proves that splitting atoms can create energy.	Jacques-Yves Cousteau and Émile Gagnan invent the Aqua-Lung.

The inventions of Jacques-Yves Cousteau helped open up a new world for exploration.

Aqua-Lung, a portable breathing apparatus that allows divers to explore underwater for long periods, in 1943. Cousteau's technology has allowed people to find out more about the oceans, their plants and animals, and the threats they face. Some consider this an achievement equal in importance to the accomplishments of space exploration.

During the last two thousand years greater scientific knowledge and more complex mathematics have helped people develop advanced inventions. Useful inventions such as telescopes and photography have helped to push science forward. In modern times, science, technology, and mathematics are equal partners, the three main tools that drive the quest for knowledge in the continuing story of human civilization.

—Chris Woodford

Further Reading

Books

Bridgman, Roger. *Eyewitness: Technology*. New York: Dorling Kindersley, 1998.

Bunch, Bryan. *The History of Science and Technology: A Browser's Guide to the Great Discoveries, Inventions, and the People Who Made Them from the Dawn of Time to Today*. New York: Houghton Mifflin, 2004.

Gribbin, John. *Science: A History: 1534–2001*. New York: Penguin, 2003.

Web sites

New York Hall of Science

A hands-on science and technology center in Queens, New York.

http://www.nyhallsci.org/

Science Museum

A museum of science and technology in London, England.

http://www.sciencemuseum.org.uk/

See also: Archimedes of Syracuse; Babbage, Charles; Cousteau, Jacques-Yves; Daguerre, Louis; Edison, Thomas; Fahrenheit, Daniel; Faraday, Michael; Fermi, Enrico; Franklin, Benjamin; Galilei, Galileo; Goddard, Robert H.; Gutenberg, Johannes; Hooke, Robert; Huygens, Christian; Lippershey, Hans; Newton, Isaac; Sturgeon, William; Volta, Alessandro.

CHRISTOPHER LATHAM SHOLES

Inventor of the typewriter

1819–1890

Modern times are often referred to as the information age because creating and exchanging information is now an important part of many people's lives. The information age gained momentum in the second half of the 20th century with such inventions as the electronic computer and the Internet. Arguably, it began a century earlier when Christopher Latham Sholes developed the first successful mechanical typewriter—a personal printing machine that enabled individuals to produce information more rapidly and professionally than they could by hand.

EARLY YEARS

Christopher Latham Sholes was born February 14, 1819, on a farm near Mooresburg, Pennsylvania. After completing school, he went to work as a printer's apprentice. Several years later, his family moved west to the newly organized state of Wisconsin. Sholes used his printing experience to enter the world of journalism. Over the next few decades, he edited a variety of newspapers in different parts of the state.

Journalism brings people into close contact with politics, an area that was of great interest to Sholes. After serving as the state printer for Wisconsin and editing the house journal of the state legislature, he served briefly as state senator. However, his health was unreliable and he lacked the robust personality needed for a political career. In his mid-forties, Sholes was appointed collector of customs taxes for the port of Milwaukee by President Abraham Lincoln. This less demanding job gave him more leisure, and he began to tinker with inventions.

Sholes had remained keenly interested in printing. In 1864, he and his friend Samuel Soulé invented a machine that could automatically print page numbers on sheets of paper. When another friend, Carlos Glidden, saw what they had done, he suggested to Sholes and Soulé that they modify their machine to also print letters. Glidden remembered seeing an article about a typing machine in the journal *Scientific American*. When Sholes read the article, he studied how the machine worked and decided he could invent something better.

MACHINES FOR WRITING

Sholes was about to invent a mechanical writing machine, but he was not the first person to try. The idea of printing pages with small metal letters (known as pieces of type) was far from new. About four hundred years earlier, around 1450, German printer Johannes Gutenberg (ca. 1400–1468) developed the modern printing press. Gutenberg's press used "movable type"—thousands of metal letters that could be rearranged in a large frame to print any number of different pages. Because

A photo of Christopher Latham Sholes from around 1880.

How a Typewriter Works

With modern computer printers, entire books can be printed in just a few minutes. Everything took much longer in the age of the typewriter, when documents had to be produced laboriously, one letter at a time, using an intricate machine built of cogs, levers, wheels, and springs.

A mechanical typewriter has several main parts. At the back of the machine, paper is wrapped around a heavy rubber roller called a platen and held in place by a curved piece of metal or plastic. A piece of inked ribbon hangs just in front of the platen, feeding sideways from a spool on one side of the machine to another spool on the other. The entire back part of the machine, including the platen, is called the carriage. It gradually moves the paper from right to left as letters are typed. At the front of the machine are rows of metal keys arranged in the familiar QWERTY layout. Each key is connected by levers and springs to a type-hammer that contains two small pieces of metal type—either an upper- and lowercase version of the same letter or numbers and characters.

When the typist presses a key, it pushes its type-hammer up toward the paper. As it does so, the ribbon rises so that it comes between the hammer's metal type and the paper. When the type hits the ribbon, an inked impression of the letter presses onto the paper. When the key is released, a spring mechanism makes the key and the type-hammer fall back down. The ribbon spools turn slightly so a fresh piece of ribbon is ready to ink the next letter. The carriage moves one place to the left so the next letter does not type on top of the first. When a whole line has been typed, a bell rings to alert the typist. Pulling on a lever (called the carriage return) moves the entire carriage back to the right and turns the paper up slightly so it is ready to type the next line.

A *vintage typewriter.*

arranging all the letters took so long, the printing press was better suited to making many copies of one page than to making one copy of a single page.

Inspired by Gutenberg's press, inventors tried to make a machine that could print single pages. In 1714, English inventor Henry Mill (ca. 1683–1771) was granted a patent for a "method for the impressing or transcribing of letters singly or progressively one after another, as in writing" that would make pages "so neat and exact as not to be distinguished from print." In 1808, Italian nobleman Pellegrino Turri built a machine so a blind countess friend of his could write her letters. Louis Braille (1809–1852), the inventor of raised writing for the blind, also invented a typewriter for blind people in the mid-19th century. American surveyor William Burt (1792–1858) was granted the first American patent for his Typographer machine in 1829. It had metal type letters mounted on a semicircular wheel. The operator turned the wheel to the right letter, pushed a lever, and the letter left its ink mark on the paper. Many similar machines appeared around this time. One of the most promising was invented by an American, Samuel Francis (1835–1886); with its black and white keys, it had the charming name Literary Piano.

> Whatever I may have felt in the early days of the typewriter, it is obviously a blessing to mankind, and especially to womankind. I am glad I had something to do with it.
>
> —Christopher Latham Sholes

THE DIFFICULTIES OF INVENTION

In 1867, working with Soulé and Glidden, Sholes developed an enormous machine about the size of a kitchen table. The three men continued to tinker with their invention, and on June 23, 1868, they were granted a patent for their "Type-Writer."

Inventing is an expensive hobby, with money required to make endless prototypes (early versions of inventions). So Sholes used one of the machines to type a letter to a Wisconsin newspaperman, James Densmore (1820–1889), who had made money inventing an oil-transportation tanker some years before. Densmore immediately agreed to invest $600 in return for a quarter of the profits. Nevertheless, he was most unimpressed with the first clumsy typewriter prototypes and urged Sholes and his colleagues to try harder.

Before long, they had come up with a machine that enabled skilled operators to type words more quickly than they could write using a pen. However, the keys on the machine were connected by levers to small hammers that pressed the type-letters against the page. The faster someone typed, the more likely these levers were to jam together as they rose and fell in quick succession. In his earliest machines, Sholes had arranged the keyboard so the letters were in alphabetical order; thus, letters that often occurred together (such as "D" and "E" in words ending in "ed") were near one another on the keyboard. When the keys were pressed quickly, their type levers tended to jam. Sholes solved the problem by rearranging commonly used letters so they were farther apart. That led him to the famous QWERTY layout (pronounced "kwerty," and named for the order of the keys in the top row of letters) that is still used in most typewriters and computer keyboards.

The typewriter was improving all the time, but whenever Sholes made a new prototype, Densmore found fault and asked him to do better. Densmore kept advancing more money and buying an ever-larger stake in the invention. Eventually, he invested over $10,000 and bought out Soulé and Glidden, becoming Sholes's main partner. The following year, 1873, the typewriting machine was as good as Sholes and Densmore could make it, and they managed to interest Remington & Sons, a gun-manufacturing company based in Ilion, New York, in producing the machine. The first Remington typewriters appeared in September of that year.

IMPROVING AN INVENTION

Early typewriters cost around $100 and were big and heavy. They also had a major drawback: they could print only in capital letters. Printing in both upper- and lowercase required two pieces of type for each letter. There were two solutions to this problem. One was to have two keys for each letter (thus entirely separate keys for "A" and "a"); some early typewriters had this format, which made them twice the size, much heavier, and more expensive. Then, in 1878, Sholes came up with the idea of adding an extra key, which he named shift, to the keyboard so that both upper- and lowercase letters could be printed from the same type hammer. Punctuation marks and symbols were added above the top row of numbers.

The typewriter sold slowly in the 1870s, but really took off a decade later. In 1888, a court stenographer from Salt Lake City, Frank E. McGurrin, invented a much faster way of using a typewriter. Known as "touch typing," it involved using more fingers to type and not looking at the keyboard so much. This helped to make typewriters more popular. As the machines themselves became faster and more reliable, they sold in ever-larger numbers—around 100,000 a year by 1900.

Christopher Sholes continued to tinker with his invention until his death in 1890. Despite the major part he had played in making the typewriter a success, he finally sold his remaining interest to James Densmore for only about $20,000. Densmore, the shrewder businessman, made an estimated $500,000 in the years that followed. When Sholes died, he was buried in an unmarked grave. In 1919, to observe

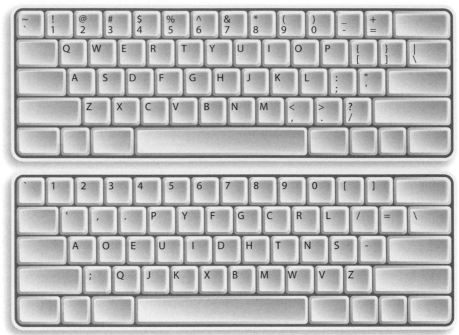

Pictured at top is the QWERTY keyboard, which was invented by Sholes and remains the most popular keyboard layout. However, it is not the only one: the Dvorak simplified keyboard, pictured in the lower half of the illustration, was patented by August Dvorak in 1936. Proponents argue that the Dvorak arrangement reduces fatigue and increases efficiency.

the centenary of his birth, the National Shorthand Reporters Association erected a more fitting memorial, engraved "In grateful memory of the man whose genius has lightened labor and brought comfort and happiness to millions of toilers in the world's work."

THE TYPEWRITER MOVES ON

Dozens of people had tinkered with typewriters before Sholes; dozens more tried to cash in on the machine's success. The earliest typewriters printed on the back of the , so it was impossible to see what was being typed without turning the paper over. "Visible" typewriters, which allowed the typist to see the letters being printed as they appeared, first went on sale in the 1880s. The rows of type keys used in early typewriters were replaced by type-wheels in some machines in the 1880s and 1890s. These wheels could be removed and swapped with different ones to change the typeface (font) being printed. The first successful portable typewriters appeared around 1910.

During the 20th century, typewriters became increasingly automated after the first electric office typewriters appeared in the 1920s. In an electric typewriter, the keys are not directly connected to the typehammers; instead, each key is a switch that operates the hammers electrically and the carriage moves back and forth electrically also. Electric typewriters are quicker and easier to use for long periods and produce a

TIME LINE

1819	1864	1868	1873	1878	1890
Christopher Latham Sholes born near Mooresburg, Pennsylvania.	Sholes and Samuel Soulé invent an automatic number-printing machine.	Sholes, Soulé, and Carlos Glidden granted a patent for a "Type-Writer."	The first Remington typewriters appear on the market.	Sholes adds the shift key to the keyboard.	Sholes dies.

more even print quality. From the 1960s onward, electric typewriters called teletypes were widely used to enter information (data) into computers and print out the results. They gradually evolved into modern computer keyboards, printers, and the word processing software that people use for writing today.

HOW TYPEWRITERS CHANGED THE WORLD

When Gutenberg developed the modern printing press, he put the power to spread knowledge into the hands of more people; when Christopher Latham Sholes invented the typewriter, that process went a step farther. Using a typewriter, one person could make professionally typed, single copies of letters, articles, and even books. A typewriter allowed anyone to set words in print and make them look important and credible.

Typewriters were also a huge force for social change. Along with telegraphs and telephones, they gave birth to the modern idea of the office: a place where information is collected and organized. Before typewriters were invented, relatively few women were working in offices. The first typing school for women was established at the New York Young Women's Christian Association (YWCA) in 1881—less than a decade after the first Remington typewriters went on sale. Within the next decade, around 60,000 women were working as typists. For some, working at a typewriter in an

Wood engraving of a woman using a Sholes type-writer, published in Scientific American *in the 1870s.*

office was not an improvement on working in the home. For others, typing skills offered a way to gain administrative skills and access to more ambitious business careers while earning their own income.

Some might argue that Sholes does not deserve credit for inventing the typewriter—many people had set the stage before he appeared on it. Yet the typewriter Sholes invented with Soulé and Glidden was the machine that won the world's heart. Historians often portray James Densmore as a greedy opportunist who swindled the inventors out of their rights to become rich from the typewriter. Yet without his investment or guidance, their machine might never have been a success. Although few people use typewriters today, most computers owe their QWERTY keyboard layout to Sholes's work; the way of building up text letter by letter and line by line on a computer screen was directly inspired by early typewriters; and computer printers also use many features borrowed from electric typewriters. The typewriter may be vanishing into history, but Sholes's inventions live on in the latest technology.

—Chris Woodford

Further Reading

Books

Adler, Michael. *Antique Typewriters: From Creed to Qwerty.* Atglen, PA: Schiffer, 1997.

Lundmark, Tobjorn. *Quirky Qwerty: A Biography of the Typewriter and Its Many Characters.* London: Penguin, 2003.

Web sites

Classic Typewriter Page
    A Web site devoted to antique typewriters.
    http://staff.xu.edu/~polt/typewriters/
Virtual Antique Typewriter Museum
    A large collection of historical photographs and documents.
    http://www.typewritermuseum.org/

See also: Braille, Louis; Communications; Gutenberg, Johannes

IGOR SIKORSKY

Inventor of the modern helicopter

1889–1972

Airplanes revolutionized long-distance travel, but they have certain limitations: they must fly very quickly to stay airborne, they need long runways for taking off and landing, and they cannot hover. A Russian engineer, Igor Sikorsky, overcame these problems when he developed the world's first practical helicopter in 1939.

EARLY YEARS

Born in Kiev, in the Russian Empire (now Ukraine) on May 25, 1889, Igor Ivanovich Sikorsky dreamed of flying even as a boy. At the age of 10, several years before Wilbur and Orville Wright made the first engine-powered airplane in 1903, Sikorsky dreamed that he was inside a luxurious, wood-paneled flying machine. Two years later, he built himself a small model helicopter powered by a rubber band. Soon he made a better one that could rise several feet into the air. Inspired by other great dreamers, he knew he would fly one day. He was captivated by the stories of French writer Jules Verne (1828–1905) and by the drawings of Italian artist Leonardo da Vinci (1452–1519), who sketched helicopter-like machines in his notebooks around 1483. Sikorsky had been born into a wealthy and well-educated family, and his parents encouraged their son's curiosity.

By the age of 14, he was ready to enter the naval academy in Saint Petersburg. Originally, Sikorsky thought he would become a naval officer, but he was much more interested in engineering. Three years later, he quit the academy to study science and math at the Kiev Polytechnic Institute.

After just one year, frustrated with academics, he abandoned his studies. During a trip to France in 1908, Sikorsky met Wilbur Wright, who was causing a sensation demonstrating his airplane to the French people. It was the first time Europeans had seen an airplane, and Sikorsky was caught up in the general mood of wonder and amazement.

HELICOPTER PIONEERS

When Sikorsky returned to Kiev, he started designing a machine with a spinning rotor that could take off straight up and thus did not need a runway. Using a rotor instead of wings was not a new idea. Chinese inventors had developed a flying top, powered by a spinning feather, around 320 CE. By 1784, two Frenchmen, M. Launoy and M. Bienvenu,

Igor Sikorsky photographed in 1942.

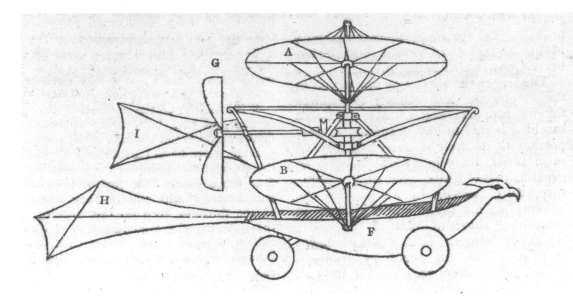

George Cayley's innovative "aerial carriage," published in Mechanics Magazine *in 1843. In theory, the circular disks (A and B) would allow the plane to fly vertically; however, the aerial carriage was never built.*

had gone one better: they built a model helicopter that flew successfully using two spinning feathers. George Cayley (1773–1857), a British air pioneer, designed a steam-powered helicopter in 1843, but it was too heavy and clumsy to fly. Finally, in 1907, two French inventors managed to lift a person off the ground in primitive helicopters. Louis Breguet (1880–1955) was first in the air: he built a four-rotor helicopter or "gyroplane" that rose about two feet (.6 m) off the ground for a minute or so. Later that year, Paul Cornu (1881–1944) flew a two-rotor machine about six feet (1.8 m) off the ground for 20 seconds.

In 1909, at age 19, Sikorsky joined the ranks of these helicopter pioneers. Borrowing money from his sister Olga, he traveled to Paris, where most of the early aviators were based. As he began meeting people, he learned what he could about building planes and bought a small airplane engine. When he returned to Kiev, he used it to build a crude two-rotor helicopter. The basic idea worked well: the rotors could generate lift (upward force), but not enough to raise the craft off the ground. The following year, Sikorsky tried a bigger engine but still failed to fly. After these two disappointments, he realized he was too short of money, experience, and materials to make a helicopter at that time.

FLYING WITH WINGS

Instead, Sikorsky decided to try making ordinary airplanes. In 1910, he built a biplane (an airplane with two wings, one above the other) called the S-1, which also failed to fly. He was more successful with his next planes—the S-2 through S-4—and by 1911 his S-5 could stay in the air more than an hour at an altitude (height above ground) of 1,480 feet

(451 m). Sikorsky's father paid for these early adventures and even mortgaged the family home to help. His faith in his son was repaid in 1912, when Sikorsky won first prize at an air exhibition in Moscow. The same year, Sikorsky won backing from the Russian Baltic Railroad Car Factory, for which he became chief aircraft engineer.

Between 1912 and the end of World War I in 1918, Sikorsky designed and built some of the best airplanes in Russia. His S-21, also known as the Grand, was a luxurious, four-engine plane, the first to have an enclosed cabin for the passengers and crew; it even had a dining table, a sofa, and a bathroom. Later, Sikorsky made a military version of the Grand that could carry bombs. This important contribution to the Russian war effort earned him an engraved gold watch presented personally by the czar (the Russian ruler).

MOVING TO AMERICA

In 1918, the Bolshevik Revolution threw Russia into turmoil. People like Sikorsky, who had become rich and famous, were targets for the Bolsheviks, who believed that money and property should be shared more equally. With a single suitcase and a small amount of money in his pocket, Sikorsky fled Russia. After visiting France briefly, he left for the United States, arriving in New York City on March 30, 1919.

Sikorsky's passport noted that he planned to build aircraft, but no one in the United States was interested—many surplus warplanes were available at discount prices. Instead he took a temporary job with the U.S. Army Air Service in Dayton, Ohio. Later he became a math teacher on New York City's Lower East Side, living in one small room

Originally built to carry passengers, this Sikorsky plane was converted into a bomber in 1915.

in Manhattan. Now age 30, he had lost his money, his fame, his homeland, and his job—but he continued to dream of building a helicopter.

Fortunately, many other Russians had fled to the United States, and some agreed to help Sikorsky restart his business. Within four years, he set up the Sikorsky Aero Engineering Company in an old barn at a farm on Roosevelt Field, Long Island. Soon renamed the Sikorsky Manufacturing Corporation, the company made successful passenger airplanes and, later, flying boats ("Clippers"), which opened up passenger routes to the Caribbean and South America for Pan-American Airways. The 1920s brought an upturn in Sikorsky's fortunes. He married his wife Elizabeth in 1924 and four years later became an American citizen. The following year, 1929, his company was taken over by the United Aircraft Corporation, which opened a large new plant in Connecticut.

BACK TO THE HELICOPTER

Although his companies in Russia and the United States had concentrated on making airplanes, Sikorsky was still thinking about how to build a helicopter. Since his early efforts in 1909 and 1910, European engineers had made great advances. In 1916, two Austrians, professor Theodor von Kármán (1881–1963) and Lieutenant Stefan Petroczy, had flown a crude helicopter to 600 feet (183 m) for an hour, though it carried no passengers and was tied to the ground with safety cables. Seven years later, Spaniard Juan de la Cierva (1896–1936) built his autogyro—a one-person airplane with small wings and a helicopter rotor on top.

If a man is in need of rescue, an airplane can come in and throw flowers on him, and that's just about all. But a direct lift aircraft could come in and save his life.

—Igor Sikorsky

Helicopters finally became practical in the 1930s. All the work that airline pioneers had done with fixed-wing airplanes made it easier for them to understand how helicopters behaved in the air. Engines and materials had also improved. Sikorsky had developed much of the theory behind the helicopter, patenting his ideas on June 27, 1931. However, he needed almost another decade to get the machine into the sky. European rivals surged ahead: in France, Louis Breguet made a two-rotor helicopter that was flying at 500 feet (152 m) by 1935; two years later, in Germany, Heinrich Focke (1890–1979) built another twin-rotor helicopter that reached altitudes of 8,000 feet (2,438 m) and stayed at that height for more than an hour. Finally, in 1939, Sikorsky started to build his own machine. Unlike the European aircraft designers, he opted for a much simpler design using a single rotor blade because, as he explained, "One person in the kitchen is fine. Two people in the kitchen get in each other's way."

How Helicopters Work

"... mankind's ancient dreams of the flying horse and the magic carpet."

A helicopter flies in a completely different way from an airplane. Its huge spinning rotor provides both its lifting power and its steering. The rotor is an immensely complicated mechanical system made up of many levers and gears, mounted on top of a powerful engine. The rotor has several spinning blades, each with a straight undersurface and a curved top edge, like the airfoil wing of an airplane. As the rotor spins, the blades chop through the air several hundred times each minute. This makes an upward force called lift that balances the helicopter's weight and keeps it in the air. As Sikorsky said, "the helicopter approaches closer than any other vehicle to fulfillment of mankind's ancient dreams of the flying horse and the magic carpet."

Using a control stick—the collective pitch—the pilot can tilt the rotor blades to a different angle (pitch) to suit different stages of flying. During takeoff, the pilot sets the blades to a steep angle so that they will generate a huge amount of lift. The lift force exceeds the craft's weight, so the helicopter rises. During landing, the pilot sets the blades to a shallow angle so that the weight exceeds the rotor lift and the helicopter lowers to the ground. For hovering in midair, the pilot sets the pitch so that the upward lift exactly balances the helicopter's weight.

Rotor blades also control how a helicopter steers. Using a second control stick called the cyclical pitch, the pilot can set the blades so they change from a steep to a shallow angle, and back again, every time they rotate. To go forward, the pilot adjusts the cyclical pitch so each blade makes a steep angle when it is at the back of the helicopter and a shallow angle at the front. The blades generate more lift at the back of the craft than at the

View of the interior of a helicopter cockpit.

front, so the nose tilts down and the helicopter flies forward. The pilot can make the helicopter fly in other directions by tilting the blades in other ways.

With only one rotor, a helicopter's entire body tends to spin around. Some helicopters have two rotors that counter-rotate (turn opposite ways) to control this spin. Sometimes the two rotors are at opposite ends of the body, as on the giant military Chinook helicopter; sometimes the rotors turn in opposite ways on the same shaft. Igor Sikorsky discovered that it was possible to build a helicopter with only one rotor provided he added a smaller rotor, mounted vertically on the tail, which balances the turning effect of the main rotor. This tail rotor may look small and unimportant, but the helicopter cannot fly safely without it.

How a Helicopter Works

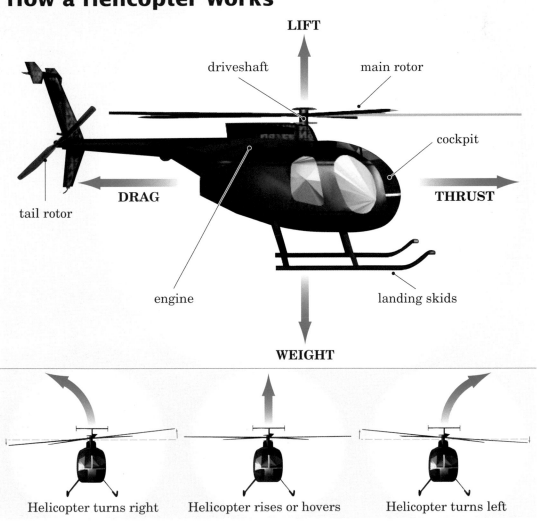

LIFT

driveshaft   main rotor

cockpit

DRAG   THRUST

tail rotor

engine   landing skids

WEIGHT

Helicopter turns right   Helicopter rises or hovers   Helicopter turns left

Sikorsky was known for always making the first test flight in his own airplanes; he was also known for his impeccable dress. Thus, on September 14, 1939, smartly turned out in a dark suit and fedora hat, Sikorsky climbed into the pilot's seat of his VS-300 helicopter, switched on the rotor, and took off. Made of welded steel tubes, with a 75-horsepower engine and a single three-blade rotor, Sikorsky's helicopter was simply built, but alarming to fly. On its maiden voyage, Sikorsky said it lurched to one side "like a frightened horse." The early days of testing saw several crashes and 19 major redesigns, but improvements came quickly. By 1940, the machine could fly for 15 minutes at a time, and in 1943 the world's first practical single-rotor helicopter went into full-scale production.

On May 6, 1941, Sikorsky set a world record, staying in the air at tree-top height for 1 hour, 32 minutes, 26 seconds. The following year, he delivered a version of his S-47 helicopter (also known as the XR-4) to the U.S. Army Air Corps. During World War II and the Korean War, Sikorsky helicopters, including the S-51 (the military version was known as R-5), proved their worth in medical, search and rescue, and combat missions. From the 1950s onward, new developments such as the jet engine (an aircraft engine powered by a mixture of hot fuel and gas roaring through it) allowed helicopters to fly farther and faster, and to carry greater loads.

Since Sikorsky's VS-300 took to the air in 1939, helicopters have repeatedly proved their worth, saving thousands of lives in famines, floods, fires, and other challenging situations. In almost every conflict

Sikorsky flies one of his helicopters in 1942.

An S-51 helicopter in 1949, with Frank Carey Sr. (right), who was the chief engineer for the device Sikorsky used to test his rotors, accompanied by his sons Frank Jr. (left) and Brian. In the background is an S-52 helicopter that just recently had broken the world helicopter speed record.

since World War II, helicopters have served as versatile attack aircraft and air ambulances. In peacetime, they have helped with everything from geological surveys to cattle ranching. Their enormous flexibility comes from their ability to take off or land vertically, hover or rotate around one spot, or drift very slowly in any direction—maneuvers no ordinary airplane can do.

LATER YEARS

Sikorsky worked for his company as aircraft designer and engineering manager. He continued to improve his helicopters and filed more patents, mainly concerned with better ways of steering and controlling his aircraft. Even after he retired in 1957 at age 68, the company kept him on as a part-time consultant. Sikorsky always commanded people's immediate respect because of his thorough knowledge of his subject. A quiet, gentle, and polite man who greeted everyone with a bow, he was said never to raise his voice and never to lose his temper.

Retirement gave him time for other interests. A refined and educated man, he wrote religious and philosophical books and helped to construct an Orthodox church for the Russian community near Bridgeport, Connecticut. He also loved the drama of the natural world: he was a keen amateur astronomer and would fly thousands of miles to

A firefighting helicopter carrying water in an emergency.

TIME LINE

1889	1909	1910	1912	1919	1923
Igor Ivanovich Sikorsky born in Kiev, Ukraine.	Sikorsky learns about building airplanes in Paris.	Sikorsky builds the S-1 through S-4 biplanes.	Sikorsky becomes chief aircraft engineer for the Russian Baltic Railroad Car Factory.	Sikorsky arrives in the United States.	Sikorsky sets up the Sikorsky Manufacturing Corporation.

watch volcanoes erupting. He died at age 83 at his family home, a farmhouse in Easton, Connecticut, on October 26, 1972.

Some of Sikorsky's success may be attributed to good fortune. Like many inventors, he quickly became bored with school and preferred to experiment independently; he was lucky to have wealthy parents who supported him. He was also fortunate in being able to immigrate to the United States. He believed that if he had stayed in Russia, his ideas would have had less chance of being developed. Nevertheless, most of his success was attributable to his technical brilliance and perseverance. Others had built helicopters before Sikorsky, but his single-rotor design proved to be the simplest and best; his basic design is still used in almost all helicopters today.

—Chris Woodford

Further Reading

Books

Hunt, William. *Helicopter: Pioneering with Igor Sikorsky.* Shrewsbury, England: Swan Hill, 1999.

Otfinoski, Steven. *Igor Sikorsky: Father of the Helicopter.* Vero Beach, FL: Rourke, 1993.

Taylor, John. *Sikorsky: Images of America.* San Francisco: Arcadia, 1998.

Web sites

How Stuff Works: How Helicopters Work
An explanation of how the various parts of a helicopter work.
http://science.howstuffworks.com/helicopter.htm

TIME LINE

1931	1939	1943	1957	1972
Sikorsky patents his theories behind the helicopter.	Sikorsky begins to build his own helicopter.	The first practical single-rotor helicopter goes into full-scale production.	Sikorsky retires.	Sikorsky dies.

How Things Fly
An interactive simulation about the basic principles of flight from NASA.
http://www.aero.hq.nasa.gov/edu/simulation.html

Igor I. Sikorsky Historical Archives
Extensive online collection of information and photographs.
http://www.sikorskyarchives.com/

Sikorsky
Home page of the Sikorsky company, now owned by United Technologies.
http://www.sikorsky.com

See also: Military and Weaponry; Transportation; Whittle, Frank; Wright, Orville, and Wilbur Wright; Zeppelin, Ferdinand.

GEORGE STEPHENSON

Pioneer of railroads

1781–1848

British engineer George Stephenson rose from poverty to become the most important figure in the early development of railroads. Before Stephenson, the fastest way of traveling across land was on horseback; by the time he died, extensive networks of railroads were being built across Europe and the United States. Railroads made trade in goods faster and cheaper, made remote regions more accessible, and provided travel opportunities beyond the wildest dreams of millions of ordinary people at the time.

EARLY YEARS

George Stephenson was born in the coal-mining village of Wylam in northern England on June 9, 1781. His father helped operate a steam engine used for pumping water out of the local coal mine. The family was poor, and neither George nor his five brothers and sisters went to school.

George grew up with a great interest in engines, became his father's assistant at age 14, and within a couple of years was promoted to a position senior to his father. Wanting to become educated, George began night classes when he was 18, only then learning to read and write. He grew into a strong, confident young man, with a reputation of being able to look after himself.

George married a local farmer's daughter, Fanny, in 1802. Their only son, Robert, also became a famous railroad engineer (see box, Like Father, Like Son). Times were hard in Britain economically and George considered moving to North America, but he persevered at home. By 1812 he had become a senior engineer, in charge of all the engines at his local coal mine. He also worked on other projects—for example, inventing a safety lamp that miners could use underground without the risk of setting off gas explosions. In 1814, he took up a new challenge, one that other local engineers had also recently attempted: he built his first railroad locomotive.

RAILROADS BEFORE STEPHENSON

Railroads had existed for centuries before George Stephenson was born. In mining areas, owners would lay wooden or iron rails across the ground so that horses could pull trucks to the nearest river or canal. One such track ran in front of George's childhood home.

The idea of building a steam locomotive to run on rails came much later. The first practical steam engine was constructed in 1712 by British inventor Thomas Newcomen; it

Portrait of George Stephenson, first published in the Illustrated London News *in 1848.*

was a huge machine used for pumping water out of mine shafts. However, these early steam engines used steam only at low pressure; thus they were not powerful enough to push themselves along in the way that a locomotive could.

High-pressure steam engines were the answer, but they were difficult to build with the technology of the time, and the danger that the water boiler in the steam engine would explode was always present. British inventor James Watt (1736–1819) who improved Newcomen's engines, thought high-pressure steam would be too dangerous. His influence delayed the development of high-pressure steam engines in Britain.

The first steam engine that moved under its own power using high-pressure steam was built in France in 1769. It was designed for road travel, rather than railroads, and was clumsy and impractical. Finally, in 1804, British engineer Richard Trevithick (1771–1833) built the first railroad locomotive; a Trevithick locomotive successfully pulled trucks at a coal mine. Trevithick's work proved that a locomotive with smooth wheels, running on smooth track, would still have enough "grip" to pull a train. However, the rails themselves tended to break under the engine's weight, and Trevithick also could not demonstrate decisively that his engines would be cheaper than horses. Discouraged, he abandoned locomotive development around 1808 and turned to other projects.

An oil painting by Terence Tenison Cuneo depicts Trevithick locomotives at a coal mine in 1804.

Soon, economic changes in Britain helped turn people's minds toward locomotives again. The country was at war with France, horses were in short supply, and the price of their fodder (food) rose. At a coal mine, on the other hand, the coal fuel for powering locomotives was virtually free. Around 1812 a few engineers at British coal mines started constructing their own locomotives, using and improving Trevithick's designs.

STEPHENSON'S RAILROADS

George Stephenson built his first locomotive in 1814. He built others, continuously improving the designs, and he soon gained a reputation as the best locomotive engineer in Britain. All his earliest locomotives, however, were slow-moving machines built for owners of coal mines. The idea of public railroads that could crisscross the country, carrying people and freight, was still distant. One or two British writers had published their ideas for a national railroad network, but they were not taken seriously at first, especially as another transportation network—the canals—had only just been created.

A painting of Stephenson (right, standing on train) examining one of his earliest trains in 1814.

What turned out to be a decisive step was taken by the merchants in the town of Darlington, in northern England, in 1818. They started planning a railroad that, although only 25 miles (40 km) in length, would be the longest in the world at that time. It would run from nearby coal mines via Darlington to the seaport town of Stockton. This was to be a public railroad that anyone could pay to use—passengers as well as commercial firms.

George Stephenson was appointed engineer to the Stockton and Darlington Railway in 1822, responsible for all aspects of the work, including laying track and building bridges. He got the directors' permission to build the track strong enough to support locomotives and horses (the original plan was to have horses pull the train cars). The challenges Stephenson faced were not just technical. A special law had to be passed

Stephenson's Locomotive

The steam locomotives that hauled mainline trains well into the 20th century all traced their ancestry back to Rocket, the engine designed and built by George Stephenson, his son Robert, and their assistants in 1829. Rocket was not the first locomotive, but it was the first fast, efficient design for one. Although later locomotives were built much bigger, they all worked in fundamentally the same way.

The basic design of a steam locomotive, whether Rocket or a later model, began with a firebox or furnace at the back, into which a driver's assistant, called a "fireman," shoveled fuel. The hot gases from the fire traveled in metal pipes called "fire tubes" through the cylindrical water boiler that formed the main body of the locomotive. Some of the heat in the gases transferred across the pipes to the water, causing it to boil and producing steam at high pressure. The gases from the fire then passed out of the fire tubes and up the chimney. Rocket was the first locomotive to have many small fire tubes, rather than just one or two large ones. This design transferred heat to the water much more efficiently, a major factor that aided its victory in the locomotive competition in 1829. Since then, all steam locomotives have been built with multiple fire tubes.

To move a locomotive such as Rocket, high-pressure steam flowed in pipes from the boiler to two hollow piston-containing cylinders, one on each side of the vehicle. The force of the steam, directed by automatic valves, pushed the pistons first one way and then the other. A rod on each piston linked to other rods that forced the "driving wheels" (the front wheels in Rocket's case) to rotate. The amount of steam reaching the cylinders and thus the speed and power of the locomotive were controlled by the driver. After leaving the cylinders, "waste" steam was directed up the chimney, which also helped draw hot gases from the fire through the fire tubes, increasing efficiency. Other features that Rocket shared with later locomotives included safety valves on the boiler, control levers to put the locomotive into reverse, and springs for a smoother ride.

A woodcut of Stephenson's Rocket, first published in 1859.

by the British Parliament to allow the line to be built, and difficult negotiations were entered into with landowners who did not want the railroad to cross their land. The line finally opened on September 27, 1825, with great ceremony and with Stephenson's engine Locomotion pulling a train of 33 cars and trucks, carrying hundreds of passengers as well as freight.

Passengers were only a minor part of the Stockton and Darlington Railway, and their cars were pulled by horses on normal days. Stephenson had now begun working as engineer on a much bigger project—the world's first mainline railroad, planned to run between the major cities of Liverpool and Manchester in England, where passengers were to be important from the beginning. Difficulties again arose with the British Parliament and with landowners, as well as with canal owners who did not want competition.

Locomotives still had to prove themselves to many doubters. For example, some directors of the new railroad thought that pulling cars on ropes using large stationary steam engines would work better. To see whether faster, better locomotives could be produced, the directors

A photograph of Rocket from 1860.

organized a trial in 1829 in which a large prize was offered for the best locomotive. This trial was easily won by Rocket, built by a firm run by George Stephenson and his son Robert and driven by George Stephenson himself (see box, Stephenson's Locomotive). The following year the Liverpool and Manchester Railway opened to the public and was a huge success. The railroad era had begun.

LATER YEARS

Now that railroads had proved successful, suddenly everyone wanted to build them, in Britain and abroad. George Stephenson had plenty of work offered to him, although nothing he did later was as important historically as his first two railroads. Stephenson had a widespread reputation as a stubborn man, and some railroad firms preferred to use other engineers, including his own former assistants.

During the 1830s, Stephenson became a rich mine owner as well as an engineer, taking advantage of new coalfields he had discovered while digging excavations for his railroads. He bought land and a large house and took up hothouse gardening, then newly fashionable. He even competed successfully with an aristocratic neighbor in attempts to grow a straight cucumber. In his last years, George Stephenson was a celebrity, touring England to give talks about his early life. He died in 1848.

THE IMPACT OF RAILROADS

It is difficult to overstate the changes that railroads brought to society. Developed in an era before automobiles, trucks, or airplanes, they amazed people with the speed at which they could transport passengers and freight. Many aspects of everyday life changed. City dwellers, for example, could choose to live in the country, traveling to work every

TIME LINE

1781	1812	1814	1822	1825	1829	1848
George Stephenson born in Wylam, England.	Stephenson becomes a senior engineer at a local coal mine.	Stephenson builds his first locomotive.	Stephenson is appointed engineer to the Stockton and Darlington Railway.	Stephenson's engine Locomotion opens the Stockton and Darlington Railway.	Stephenson's Rocket wins a prize for best locomotive.	Stephenson dies.

day by train. Railroads also opened up remote areas to new human settlement.

The first locomotive in the United States was imported from Britain in 1829, but domestically built models soon became commonplace. The U.S. railroad network expanded rapidly. The Civil War (1861–1865) has been described as the first railroad war, with rail networks playing an important role in transporting troops and supplies. A landmark came in 1869, when railroads were built from the west and the east, meeting in Utah to complete the first transcontinental line. By the end of the 19th century, more than 200,000 miles (321,867 km) of railroad track wove across the United States.

Like Father, Like Son

George Stephenson's son Robert (1803–1859) was only 19 when he began to run the locomotive works established jointly by his father and local businessmen. Better educated than his father, Robert was also different in character—known among his peers as being more willing to give credit to other people. Robert and his father worked well together, however, and later Robert supervised large railroad projects of his own.

Most notably, Robert pioneered large and daring designs for railroad bridges, including a massive bridge over the Saint Lawrence River at Montreal. Although George Stephenson is better known in modern times, Robert was quite famous during his lifetime and was given the rare honor (for an engineer) of being buried in London's Westminster Abbey.

A portrait of Robert Stephenson from the Illustrated London News.

Although trains have declined in importance as a method for moving people, they remain vital for moving freight of all kinds.

Steam locomotives remained the main power units on the world's railroads until around the middle of the 20th century. (The world's largest locomotives, the Union Pacific's "Big Boys," were built as late as the 1940s.) Though powerful, steam locomotives needed careful maintenance and were very wasteful of fuel. Contemporary locomotives are either diesel-powered or electric, and thus are more efficient.

Faced by competition from road and air travel, railroads declined in overall importance later in the 20th century, especially in the United States. However, they still have important advantages. For example, mass-transportation rail systems in cities help avoid the gridlock that road travel tends to cause. Railroads also have far fewer accidents per passenger than do highways. Railroads also help the environment: they take up much less land, and less fuel is burned than in transporting the same number of people by automobile.

—Richard Beatty

Further Reading

Bailey, Michael R., and John P. Glithero. *The Stephensons' Rocket: A History of a Pioneering Locomotive*. East Lansing: Michigan State University Press, 2003.

Herbst, Judith. *The History of Transportation*. Breckenridge, CO: Twenty-First Century Books, 2005.

See also: Newcomen, Thomas; Transportation; Woods, Granville.

LEVI STRAUSS

Inventor of Levi's blue jeans

1829–1902

Nearly every man, woman, and child has a pair of jeans in the closet. Wearing jeans has become such a way of life in the United States and around the world that imagining a time without them may be difficult. The original jeans debuted in 1873 and are still in production more than 130 years later. Initially manufactured as work clothes, jeans became fashionable in the early 20th century, coveted as a symbol of the American West, and, later, of youth rebellion and counterculture. Levi's, the brand that started it all, is named after Levi Strauss, the San Francisco merchant. He did not invent jeans himself but did set jeans on the path to becoming an American icon.

EARLY YEARS

Levi Strauss was born Loeb Strauss in 1829, in what is now Buttenheim, Germany. After his father died of tuberculosis, and because of the growing poverty and anti-Semitism in Bavaria at the time, he moved to the United States with his mother and two sisters in 1847. They joined two half brothers who had already moved to the Lower East Side of New York City to set up a dry goods business. J. Strauss Brothers and Co. sold, among other items, canvas, cloth, linen, and clothing; Strauss's father had sold cloth and other dry goods door-to-door in Buttenheim.

After three years in America, working for his brothers' stores, Loeb changed his name to Levi. In January 1853, he became an American citizen and headed west to San Francisco to establish the West Coast outpost of the family business.

San Francisco was still caught up in the economic boom ushered in by the gold rush of the late 1840s. Strauss, just 24 years old, arrived in March 1853 and set up a wholesale dry goods business, Levi Strauss and Co., near the city's waterfront.

A TWO-MAN JOB

Strauss developed a reputation as an honest and fair merchant, and Levi Strauss and Co. grew rapidly into a successful business. Soon, Levi Strauss and Co. was selling cloth, blankets, handkerchiefs, and other goods to small stores throughout the West.

In 1872, Strauss received a letter from Jacob Davis, a wholesale customer from Reno, Nevada. Davis, a Latvian-born tailor, fashioned clothing out of the fabric he purchased from Strauss. Davis had a customer who had commissioned him to make reinforced workpants, ones that would not split at the seams. (According to one account, the customer was the wife of an overweight woodcutter.) Davis obliged, hammering copper rivets into the seams that were stressed during wear, including the corners of the pockets and the base of the fly.

In 2003, in honor of the 150th anniversary of the founding of Levi Strauss and Co., the firm established a visitors' center featuring vintage advertisements and memorabilia at its headquarters in San Francisco.

Davis's riveted work pants were a resounding success. Word of mouth had their popularity soaring. Sensing a major business opportunity, and fearing imitators, Davis wrote to Levi Strauss. He wanted to patent the riveted pants, but lacked the funds to pay the $68 patent fee. If Strauss could pay the fee, Davis offered, they could produce the riveted workpants together. He assured Strauss that they would make a very large amount of money. On May 20, 1873, Strauss and Davis received U.S. Patent No. 139,121 for their "Improvement in Fastening Pocket-Openings." This day is widely considered to be the birthday of blue jeans.

THE JEANS BUSINESS

Although Jacob Davis may have technically invented riveted work pants, Levi Strauss's business acumen ensured their success. Strauss brought Davis to San Francisco to oversee the production of the first copper-riveted "waist overalls" (the original name for jeans). Originally, the pants came in two styles: one made from cotton duck, a heavy brown cloth; and another from a sturdy blue denim. Denim later became the standard. At first, the cloth was cut and sent to seamstresses who worked from home, but as demand grew, Strauss built factories throughout San Francisco.

The heavy-duty, high-quality, durable construction made the pants perfect work wear for miners and other hard laborers. The riveted waist overalls revolutionized working clothes, becoming so popular that they came to be known simply as Levi's.

When the patent expired in 1891, dozens of imitators began making similar cltohes. True Levi's could be identified by the now famous lot number "501," the double arc stitching on the pockets, and the leather waistband patch bearing the two-horse logo. These details were all added to discourage counterfeiting. Each of these elements is still used.

Levi Strauss and Co. soon branched out into jackets, outerwear, and shirts. Toward the end of the 19th century, Strauss himself passed the day-to-day responsibilities to younger family members and turned his energies to other civic and business activities. Strauss was very active in San Francisco's Jewish community and was a member of San Francisco's first synagogue, Temple Emanu-El. As a longtime member of the San Francisco Board of Trade and as director of, variously, a bank, an insurance company, and the San Francisco Gas and Electric Company, he was also deeply engaged with the business community. His philanthropy was well known—in addition to funding more than two dozen scholarships at the University of California–Berkeley, Strauss contributed to many charities and local orphanages.

In 1902, Strauss died, leaving the bulk of the business to his nephews. He was remembered as one of San Francisco's leading citizens.

LEVI'S AFTER LEVI

The company headquarters and factories were destroyed during the devastating San Francisco earthquake in 1906. Levi Strauss and Co. rebuilt and by the 1920s had become the leading brand of men's work wear in the western United States. In the 1930s and 1940s, with the popularity of early westerns on movie screens, Levi's became a symbol of the

Denim vs. Jeans

Much debate swirls around the origins of the word *jeans*. Some scholars say that *jean* comes from the word for a cotton fabric used to produce men's pants in Genoa, Italy, in the 1600s—*jean* being an English abbreviation of *Genoa*.

By the end of the 1700s, two all-cotton fabrics were being produced in America, one called "denim" and the other called "jean." Denim (supposedly named after the city in France where it was first made: de Nîmes, meaning "of Nimes") was heavier and more durable. This is what Levi Strauss used for his riveted overalls in the late 1800s. Denim was a mix of white yarn and dyed yarn, usually an indigo color. Jean was produced with one color of thread and was not as rugged.

Jean cloth declined in popularity in the 19th century, just as denim grew in popularity because of Levi's. The word *jean* resurfaced in the 1950s and 1960s. It was adopted, first by the public and then officially by Levi Strauss and Co., as the accepted name for blue denim pants.

Actor James Dean helped solidify blue jeans' iconic status in films like Rebel without a Cause *(1955).*

American West, popularized by celebrities such as John Wayne, Gene Autry, and Gary Cooper. Levi Strauss and Co. capitalized on this imagery by advertising its "authentic cowboy pants." The pants were slowly becoming recognized more as leisure wear and less as work wear.

Levi's were introduced to Europe in the 1940s by soldiers stationed abroad during World War II. By the 1950s, the pants were sold in the Midwest and the East for the first time. Postwar teenagers and young people began to wear Levi's, emulating the rebellious look and spirit of screen stars like Marlon Brando and James Dean.

By the 1960s, the original riveted-waist overalls had evolved into the jeans we know today: belt loops were added in 1922; the zipper replaced the button-fly in 1955; and rivets disappeared from the back pockets in 1967.

Levi's enjoyed an even more stunning reputation abroad, particularly in the Soviet Union and in communist eastern Europe, where jeans became a highly coveted black market item in the 1980s. Indeed, the simple pair of denim pants came to represent American ideals of freedom, equality, and rugged individualism.

THE BRAND LIVES ON

Although invented in the United States, jeans have become a global wardrobe staple.

Amidst the modern deluge of denim brands, Levi's remain one of the most popular brands of jeans in the world. In 1994, *Fortune* magazine named Levi Strauss and Co. the most admired apparel company in the United States, not just for its jeans but also for its continuing spirit of philanthropy inspired by Levi Strauss. The company was an early leader

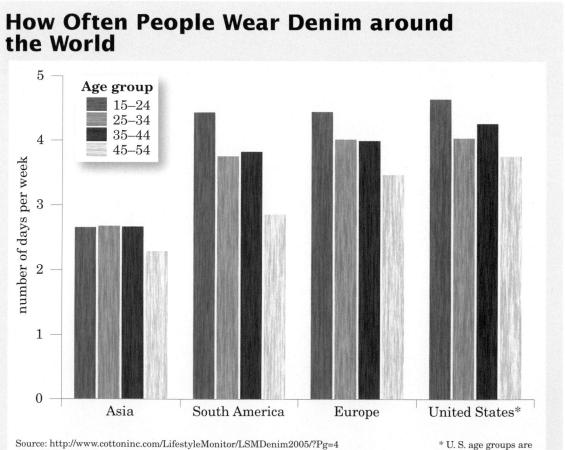

How Often People Wear Denim around the World

number of days per week

Age group
- 15–24
- 25–34
- 35–44
- 45–54

Asia — South America — Europe — United States*

Source: http://www.cottoninc.com/LifestyleMonitor/LSMDenim2005/?Pg=4
Global Lifestyle Monitor III, 2003 (Cotton Incorporated and Cotton Council International).

* U. S. age groups are 16–24 and 45–55

TIME LINE

1829	1847	1850	1853
Levi Strauss born as Loeb Strauss in Buttenheim, Germany.	Strauss moves to the United States.	Strauss changes his first name to Levi.	Strauss moves to San Francisco and founds Levi Strauss and Co.

in establishing AIDS support groups for employees (1983); in offering full medical care benefits to the unmarried partners of employees (1992); and in adopting worldwide standards for all its contractors regarding wages, hours, working conditions, and environmental responsibility (1993). In addition, Levi's are hailed as one of the few products to remain essentially the same since they were first produced. In the words of Kenny Wilson, a company executive, "There are very few categories of product in the world that are made perfect the first time around."

Levi's can be found in closets throughout the world, as well as at the Smithsonian Institution in Washington, D.C. This is a testament to the ingenuity of Jacob Davis and Levi Strauss, two 19th-century European immigrants who saw a need and met it with simple copper rivets and good salesmanship. Levi Strauss did not live to see his waist overalls become one of the most enduring and quintessential symbols of

Shopping for Levi's in Hong Kong, in 2001.

1872	1873	1891	1902
Strauss receives a letter from Jacob Davis asking for help in obtaining a patent for his work pants.	Strauss and Davis receive a patent for their riveted work pants.	Strauss and Davis's patent expires.	Strauss dies.

Levi's for women were first featured in *Vogue* magazine in 1935.

America, but Levi Strauss and Co., now more than 150 years old, has seen generation upon generation of Americans embrace its founders' most popular goods, jeans.

—Laura Lambert

Further Reading

Article
Davis, William A. "Levi Strauss Museum Gives the 411 behind the 501s." *Boston Globe*, November 19, 2003.

Books
Cray, Ed. *Levi's*. Boston: Houghton Mifflin, 1978.
Marsh, Graham. *Denim: From Cowboys to Catwalks*. London: Aurum, 2002.

Web sites
Lemelson-MIT Program: Inventor of the Week—Levi Strauss
　　A profile of Strauss.
　　http://web.mit.edu/invent/iow/strauss.html
Levi Strauss and Co.
　　Official history from the company.
　　http://www.levistrauss.com/about/history/founder.htm

See also: Cloth and Apparel.

WILLIAM STURGEON

Inventor of the electromagnet and the commutator

1783–1850

During the 19th century, electricity went from being a mysterious natural phenomenon to being the basis of an international engineering industry that provided light, heat, and communications for the industrialized world. A key figure in that shift was the British electrical engineer William Sturgeon, who developed two crucial inventions that helped harness electricity for useful work.

EARLY YEARS

William Sturgeon was born in Whittington, Lancashire, in northern England, on May 22, 1783. He was the son of a shoemaker, but both his parents died while he was still young, and he was apprenticed to another local shoemaker at the age of 13. After six years he joined the army: first the local Westmorland Militia, and then, in 1804, the Royal Artillery, based at Woolwich in south London. He had little formal education, but his natural curiosity about science was noticed and encouraged by various officers there, who lent him books so he could pursue his growing interest in physics. Sturgeon became increasingly interested in the phenomenon of electricity, which had caught the popular imagination at that time. He also developed a reputation as a skilled public demonstrator of electrical apparatuses that he built himself—mostly novelty items designed to give mild electric shocks.

When Sturgeon left the army in 1820, he combined his earlier trade with his army connections by opening a boot-making business in Woolwich. However, he soon decided to concentrate his skills on making scientific instruments. He also tried to earn a living by giving public demonstrations about science. His skills as instrument maker and showman led to an appointment as a science teacher at the Military Seminary in Addiscombe, Surrey, in 1824.

THE ELECTROMAGNET

In 1820 Sturgeon had been excited to hear about a discovery by Hans Christian Ørsted (1771–1851), a Danish scientist. Ørsted found that a compass needle, when brought close to a wire through which an electric charge was flowing, would be deflected from the magnetic North Pole. When the current was switched off, the needle resumed its original position. Ørsted concluded that an electric current created a magnetic field, signifying that electricity and magnetism, which had previously been considered separate forces, must be related.

Sturgeon realized that a strong current could be used to produce a powerful magnet. He coiled an electric wire tightly

Engraving of William Sturgeon from around 1830.

Who Invented the Electric Motor?

British physicist Michael Faraday (1791–1867) took the first steps toward creating motion from electricity with an experiment in 1821. He suspended a wire from a hook, with the free end in a pan of mercury (a good electrical conductor), in the middle of which was a fixed magnet. When the pan and the wire were connected to a battery to form a circuit, the free end of the wire moved in a circle around the magnet. This showed that magnetic field lines were perpendicular to the electrical source that produced them and moved in a circle around the source.

Ten years later, American physicist Joseph Henry (1797–1878) built a device in which an electromagnet balanced like a seesaw on a pivot, with a permanent magnet under each end. At the bottom of its swing, each end in turn made contact with a battery, completing an electrical circuit that gave it a polarity identical to the magnet beneath it. As a result, the beam was repelled upward. At the bottom of the next stroke, the circuit was completed on the other side, with the opposite polarity, repelling the beam again. This process created a continuous rocking motion.

In Sturgeon's design, each magnet created its own magnetic field, and the freely moving magnet turned on a spindle until it aligned with the stationary magnet. The rotation of the spindle caused each of the wires from the battery to come into contact with the opposite commutator plate. This immediately reversed the direction of current in the circuit, making what had been the north pole of the moving magnet into the south pole, and vice versa. The stationary magnet, its polarity unchanged, now repelled the moving one, causing the spindle to make a further half turn. The direction of current reversed itself each half turn, causing the spindle to rotate continuously.

Why was Sturgeon's commutator such a breakthrough? Because once a way had been found to make a power source produce rotary motion, it could be used to power any type of machinery. That had happened earlier, when the steam engine was transformed from a beam engine, which could only pump water, to a wheeled engine, which could power textile mills and railroad locomotives. It would happen again later when the internal combustion engine was attached to a crankshaft, enabling it to power automobiles and airplanes.

William Sturgeon's Electric Motor

battery   spindle   metal plates (commutator)   rotating magnet   stationary magnet   battery

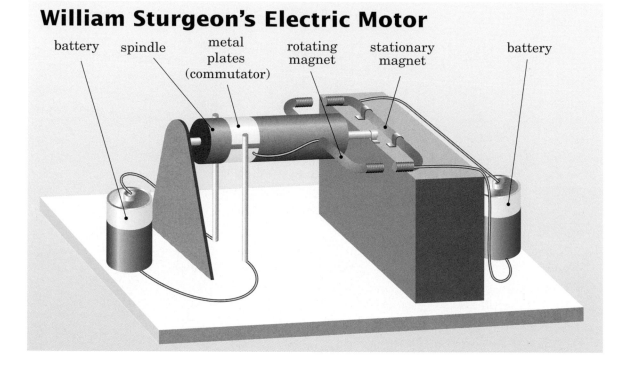

Sturgeon's commutator was revolutionary because it used a power source (batteries) to produce rotary motion. This eventually led to the development of other types of electric motors.

around an iron bar, creating what is known as a solenoid. The coils greatly strengthened the magnetic force, since each coil's magnetic field reinforced those around it. The iron bar became, in effect, a giant compass needle, with one end of it a north-seeking pole, and the other a south-seeking pole.

Knowing that opposite or unlike poles attract each other, whereas like poles repel each other, Sturgeon then had the idea of bending the iron bar into a horseshoe shape to bring the opposite poles next to each other. The strong attraction between them concentrated the magnetic field into a smaller space, making it more powerful. In 1825 Sturgeon gave a public demonstration of a horseshoe-shaped electromagnet weighing just 7 ounces (200 g): when the current from a battery was passed through the solenoid, it was able to lift metal weighing 9 pounds (4 kg), or 20 times the magnet's own weight. In the contemporary world, huge electromagnets capable of lifting many tons are used in steelmaking, mining, and shipbuilding.

THE COMMUTATOR

The ability of electricity and magnetism to deflect a compass needle made many people think that there must be a way to harness electricity to create motion. Like many great advances in technology, the development of the electric motor was the result of many individuals' contribution of ideas (see Box, Who Invented the Electric Motor?).

Sturgeon began working on the problem in 1832. He used two horseshoe-shaped electromagnets, one of which was fixed while the other, positioned across from it, was mounted on a spindle so it could turn. His innovation was the commutator, a pair of metal plates wrapped around the spindle through which the current passed from the battery to the magnet—thus creating a circuit. As the spindle turned, the commutator switched the direction of the electric current every half turn, so that the stationary magnet alternately attracted and repelled the moving magnet, causing the spindle to rotate continuously.

The results of Sturgeon's invention were far-reaching. Apart from all the uses of an electric motor, the idea could also be put into reverse. If one turned the spindle to move the magnets, it became a generator producing alternating current (AC) electricity. Whatever their ultimate power source, all power stations now produce electricity by turning a spindle with a commutator that rotates through a magnetic field.

LATER YEARS

In the same year he invented the commutator, Sturgeon took up a post demonstrating his electrical instruments at the Adelaide Gallery of Practical Science in London, a sort of museum that mixed entertainment with education and survived on donations from the public. In 1837, hoping to generate more debate about new advances in both the

In 2006 workers installed an electromotor commutator at a hydropower station at the Three Gorges Dam in Yichang, China.

TIME LINE

1783	1804	1820	1824	1825
William Sturgeon born in Whittington, England.	Sturgeon joins the Royal Artillery.	Sturgeon leaves the army and opens a boot-making business.	Sturgeon becomes a science teacher at the Military Seminary.	Sturgeon demonstrates a horseshoe electromagnet.

theory and the technology of electricity, he formed the Electrical Society of London and edited its publication, the *Annals of Electricity*, until its last issue in 1843.

By that time he had moved to Manchester, where in 1840, following the Adelaide Gallery's closure that year because of lack of funds, he

The Romance of Electricity

In the late 1700s and early 1800s, the artistic and political movement known as romanticism was sweeping Europe and North America. One of its hallmarks was a sense of awe at the mysterious powers of nature. In 1791 the Italian anatomist Luigi Galvani (1737–1798) was dissecting a frog when he accidentally touched it with a scalpel that had built up a small static charge. The frog's leg twitched, even though it was dead. Writers and artists wondered: What kind of life-giving properties might electricity have if it could make a dead body move? In 1818, just before Sturgeon began his experimental work, Mary Shelley's novel *Frankenstein* imagined a creature that was sewn together from parts of dead bodies and brought to life by a strong charge of electricity from a bolt of lightning.

By the time the British physicist James Clerk Maxwell (1831–1879) delineated the relationship between electricity and magnetism in a set of four short equations in 1873, electricity was no longer an unknowable or untamable force. Nevertheless, in addition to generating power for lightbulbs, it could still generate a sense of wonder.

TIME LINE

1832	1837	1840	1847	1850
Sturgeon invents the commutator.	Sturgeon founds the Electrical Society of London.	Sturgeon appointed superintendent of the Royal Victoria Gallery of Practical Science.	Sturgeon is granted a pension by the British government.	Sturgeon dies.

had been appointed superintendent of a similar venture, the Royal Victoria Gallery of Practical Science. After four years the gallery met the same fate as its London counterpart, and Sturgeon held no regular post after 1844. His friends lobbied for him to receive a state pension, which was granted in 1847, though his life was always one of financial struggle; he had never patented any of his inventions, believing they should be freely available to benefit everyone. He died on December 4, 1850, in Prestwich, Manchester.

At a time when those without a university degree had difficulty being taken seriously by the scientific community, Sturgeon managed to impress his peers with the inventiveness of his devices, as well as the depth of his theoretical understanding. The importance of his legacy has grown ever since, and in the 20th century his inventions became the basis for industrial processes and electricity generation all over the world.

—Jonathan Dore

Further Reading

Books

Gutnik, M. J. *Electricity from Faraday to Solar Generators*. New York: Franklin Watts, 1986.

Hart-Davis, Adam. *100 Local Heroes*. Stroud, Gloucestershire, UK: Sutton, 1999.

Porter, Roy. "William Sturgeon." In *The Biographical Dictionary of Scientists,* edited by Roy Porter. New York and Oxford: Oxford University Press, 1994.

Web sites

Development of the Electric Motor
    Spark Museum's page, with many photographs of early electric motors.
    http://www.sparkmuseum.com/motors.htm

Famous Scientists
    Web pages on many of the most important electrical engineers and scientists of
    the time, including Sturgeon and his contemporaries.
    http://chem.ch.huji.ac.il/~eugeniik/history/electrochemists2.htm
William Sturgeon, 1783–1850
    Adventures in Cybersound page with three biographical articles about Sturgeon
    plus further reading.
    http://www.acmi.net.au/AIC/STURGEON_BIO.html

See also: Energy and Power; Faraday, Michael; Tesla, Nikola; Volta, Alessandro.

MARIA TELKES

Inventor of solar technologies

1900–1995

Maria Telkes was a pioneer in the development of solar energy technology. She created useful solar-powered ovens and water purifiers, and she experimented with solar technologies for the home. Although far from the first to exploit the sun as a power source, Telkes proved that harnessing solar energy, an idea that was considered somewhat fantastical, could be made practical.

EARLY YEARS

Maria Telkes was born on December 12, 1900, in Budapest, Hungary. Telkes became fascinated by solar energy as a teenager, when she read a book on the future of power. She went on to read all the material she could find on the subject, but at the time, solar energy was used only haphazardly and little in the way of technology was specifically designed to tap the power of the sun.

A strong student, Telkes attended the University of Budapest, where she earned a bachelor's degree in 1920 and a doctorate in physical chemistry in 1924. The next year, she came to the United States to visit a relative living in Cleveland, Ohio.

While in Ohio, Telkes received a job offer from the Cleveland Clinic Foundation, a hospital that was looking for a biophysicist to help conduct medical research. Telkes accepted the position, working for the Cleveland Clinic until 1937, the year she became a U.S. citizen. That year, she moved to East Pittsburgh, Pennsylvania, and began working for the Westinghouse Electrical and Manufacturing Company as a research engineer.

SOLAR HOMES

Telkes was still interested in solar energy, but it was considered so unlikely to ever become practical that ongoing research was minuscule and the field offered few job opportunities. Nonetheless, solar energy did have its proponents. During the 1930s, the United States and the world were in a severe economic recession, and many people could not afford fuel to heat their homes in winter. The idea of using sunlight to heat homes became more and more popular, leading to a minor boom in solar homes. Solar homes did not use modern solar technology like photovoltaic panels; instead, the homes

Maria Telkes, photographed in 1953.

were designed with windows and skylights that captured the sun's heat.

Although well-designed solar homes could reduce heating costs, a traditional furnace was still required. The need for better solar homes eventually led the Massachusetts Institute of Technology (MIT) to initiate a program of solar research in 1937. Two years later, Telkes was finally able to pursue her interest in solar power when she was hired by MIT as a research associate.

SOLAR WATER

Events soon altered the course of her research. The United States entered World War II in 1941, and all the scientists in MIT's solar program, including Telkes, were reassigned to war-related projects.

During the war, many of the battles fought between the United States and Japan took place in the tropical latitudes of the Pacific Ocean. Ships and planes were shot down frequently, and pilots and sailors were sometimes stranded on life rafts under the blazing sun for days at a time while awaiting rescue. Dehydration was a serious threat to survival (seawater contains too much salt to drink).

The military needed a portable water-purification kit that could function in the most primitive conditions, so Telkes began investigating solar water purifiers. Such purifiers worked by capturing the heat of the sun and using it to evaporate water. The water vapor condensed in a cooler part of the device, and the liquid was collected. As salt did not evaporate like water, the collected water was pure and fresh.

Solar water purifiers already existed, but they were far too large and heavy to be part of a portable emergency kit. Telkes realized that she could make a lightweight, collapsible purifier using clear plastic film. It eventually became standard gear in military emergency kits, saving countless lives.

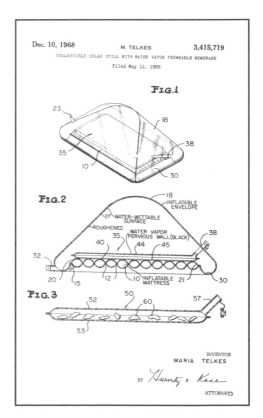

Illustration from one of Telkes's patents: a portable solar water purifier or, as she described it here, a collapsible solar water still.

SOLAR COOKING

Solar ovens are designed to gather the maximum amount of sunlight possible. The reflector aims sunlight into the chamber, which is colored black in order to absorb the heat. Insulated walls and a glass cover help trap the heat inside the oven.

After World War II ended in 1945, Telkes and other scientists at MIT returned to working on solar houses. The solar homes built by MIT's researchers were typically designed to collect and store solar energy during the day and to release it at night. Telkes decided that heat storage could be taken much farther, and she created a house that could be kept warm all winter using only energy from the sun. The house was constructed in Dover, Massachusetts, in 1948 (see box, The Dover Sun House).

In 1953, Telkes took a job at New York University in New York City. The next year, she received a grant from the Ford Foundation to develop a solar oven for use in poor countries. Earlier designers had created solar ovens that used specially crafted curved mirrors to focus the sun's energy and to generate enough heat to cook food, but these mirrors were expensive. Telkes's solar oven worked according to a similar principle, but it was simple enough in design to be made at home using basic tools. It could also use anything reflective, including flat mirrors, metal, or foil.

How a Basic Solar Oven Works

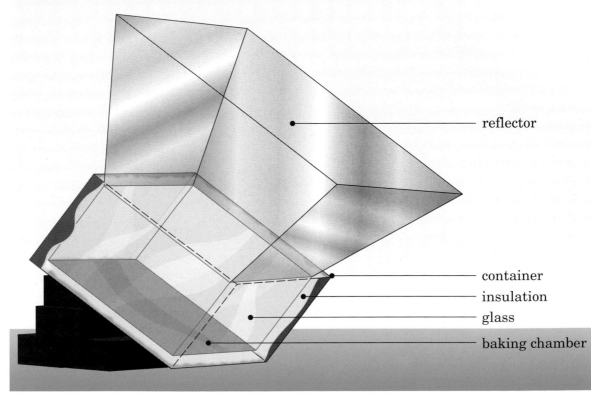

reflector

container
insulation
glass
baking chamber

LATER YEARS

Telkes moved away from academia and solar energy research during the 1960s, taking corporate jobs that focused on packaging designed to keep contents at a specific temperature—technology that was derived from the work she did on the Dover solar house. In the 1970s, she became a researcher at the University of Delaware, again developing technology to make solar homes more practical.

Telkes retired in 1977, but she continued to work as a consultant for the university until the early 1990s. In 1995, Telkes, who was living in North Miami, Florida, decided to visit Hungary for the first time in 70 years. She died in Budapest on December 2, 1995, at the age of 95.

Telkes's work helped spur a conceptual breakthrough about solar power, demonstrating that the energy produced by the sun could be tapped in practical, useful ways. Telkes's water purifier and stove are still used, and although most homes still employ conventional heating and

The Dover Sun House

Whereas earlier designs had relied on water or stone to store heat, Telkes came up with a novel storage material: Glauber's salt. This salt, used in textile dyeing, melts at a fairly low temperature. Once melted, it can absorb a great deal of heat, which it releases as it refreezes. Telkes determined that a tank of Glauber's salt could store enough heat to keep a house warm for several winter days even with a completely overcast sky.

MIT would not support the development of the unusual idea, however, so Telkes found outside funding. In December 1948, a house was completed in the chilly climate of Dover, Massachusetts, that had no furnace, but was heated by solar energy stored in Glauber's salt. The system could also be used to cool the house in the summer. The solar heating system worked relatively well, although it was discovered that an 11-day stretch of cloudy days would exhaust the heat-storage system.

After three years the salt deteriorated, and the experimental solar system was eventually replaced with a conventional furnace. Nonetheless, during those three years the furnace-free Dover Sun House was an object of intense media scrutiny, spurring popular interest in solar energy.

TIME LINE

1900	1924	1944	1948	1954	1977	1995
Maria Telkes born in Budapest, Hungary.	Telkes earns a doctorate in physical chemistry from the University of Budapest.	Telkes invents a portable solar water purifier.	The Dover Sun House is built.	Telkes receives a grant to develop her solar oven.	Telkes retires.	Telkes dies.

electrical systems, tens of thousands of homes in the United States now rely partly or wholly on solar power.

—Mary Sisson

Further Reading

Books

Behrman, Daniel. *Solar Energy: The Awakening Science.* Boston: Little, Brown, 1976.

Butti, Ken, and John Perlin. *A Golden Thread: 2,500 Years of Solar Architecture and Technology.* Palo Alto, CA: Cheshire, 1980.

Web sites

Architects & Buildings: Solar Houses
  Information on the solar architecture project at MIT.
  http://libraries.mit.edu/guides/subjects/architecture/architects/solar/index.html
Maria Telkes: Solar-Powered Devices
  A profile of Telkes by the Lemelson-MIT Program.
  http://web.mit.edu/invent/iow/telkes.html

See also: Energy and Power; Ovshinsky, Stanford.

EDWARD TELLER

Inventor of the hydrogen bomb

1908–2003

Edward Teller developed the hydrogen bomb, an extremely powerful and deadly weapon that generated tremendous political controversy and also ushered in what came to be called the nuclear age. Teller was a strong advocate for nuclear technology. As a result, he became a highly divisive figure among scientists and in the larger world.

EARLY YEARS

Edward Teller was born in 1908 in Budapest, Hungary. His father was a lawyer, and the Tellers were fairly wealthy, though they experienced significant hardship when the Communist Party took control of Hungary in 1919: the party put all commerce and industry under government control, closing down private markets. The economy failed, and city dwellers like the Tellers who did not grow their own food could not find it for sale. "I have clear memories of hunger during that period," Teller later wrote.

Despite these hardships, Teller received a good education. He was fascinated by math and planned to study it in college. His father, however, worried that Teller would never get a job with a math degree, so Teller agreed to study chemistry.

In the fall of 1925, Teller enrolled in the Institute of Technology in Budapest. He stayed for only a few months, however, because his father decided that the anti-Semitic policies of the Hungarian government would prevent Teller, who was Jewish, from finding work after he graduated. In 1926, Teller went to college in Germany, which at the time was believed to be more hospitable to Jews.

PHYSICS AND HITLER

Teller then enrolled in the Karlsruhe Technical Institute in Karlsruhe, Germany, where he studied chemistry. There he heard a great deal about physics, which at the time was a rapidly developing field. Teller became so interested in physics that he decided to switch fields, and in 1928 he matriculated at the University of Munich because it had a stronger physics program.

That summer Teller's foot was run over by a streetcar and had to be amputated. He went back to Budapest to convalesce and then returned to Germany, this time attending the University of Leipzig to study under the famous

Edward Teller in 1963.

physicist Werner Heisenberg (1901–1976). In 1930, Teller received his doctorate from Leipzig as well as an offer to work at the university as Heisenberg's assistant. A year later, he received a job offer to be a research consultant at the University of Göttingen.

Teller was working at Göttingen when Adolf Hitler came to power in January 1933. Teller was far more interested in physics than in politics, but he knew that Hitler's Nazi Party was extremely anti-Semitic. Although Teller did not realize the danger he was in (the Nazis would systematically execute six million Jews across Europe), he knew that job prospects for Jews would disappear and he decided to leave Germany.

Teller took various positions at universities in Copenhagen and in London. In 1935, he accepted an offer to become a full professor at George Washington University in Washington, D.C.

THE ATOMIC BOMB

Teller was in Washington, D.C., in early 1939, when a stunning announcement was made: German physicists had split the atom. This process, called fission, released a great deal of energy. To a physicist like Teller, the implication was clear—Nazi Germany had discovered something that could be used to make an incredibly powerful bomb.

To those who were not physicists, however, the implication of the German experiment was not as obvious. Teller became involved with a group of physicists who lobbied the U.S. government to fund a program that would build a fission bomb—better known as an atomic bomb—before the Nazis. Under Nazi rule, Germany began invading and conquering other countries in Europe; the United States began a program to develop an atomic bomb.

Teller became a U.S. citizen in March 1941; shortly thereafter, he was invited to join the atomic bomb program. The United States would enter World War II in December of that year, following the Japanese bombing of Pearl Harbor in Hawaii. Teller moved to New York City and then to Chicago to work on the bomb.

While in New York, Teller had a conversation with physicist Enrico Fermi (1901–1954) about the possibility of using an atomic bomb to set off another kind of reaction—a fusion reaction. In a fission reaction, a large atom such as uranium or plutonium is split; in a fusion reaction, smaller atoms are fused together, a process that releases even more energy than fission (fusion is the process that fuels the sun and stars). Teller was immediately intrigued by the potential power of a fusion bomb, also called a hydrogen bomb because isotopes of hydrogen undergo fusion.

Compression and Reaction

Atomic bombs explode when the plutonium or uranium in them undergoes a fission reaction. A fission reaction, however, will happen only if a large enough amount of plutonium or uranium is gathered together in one place. The amount of material that will cause an explosion is called a critical mass.

During the development of the atomic bomb in the 1940s, one of the challenges was to develop a trigger, a device to set off the fission reaction. Two different kinds of triggers were used. One kind joined two chunks of uranium to make a critical mass. That trigger was inappropriate for plutonium, however, which is less stable than uranium: the danger was that the two chunks of plutonium would react with each other and undergo fission before they were supposed to.

Teller helped design a second kind of trigger—an implosion trigger. He realized that if a small amount of plutonium, less than a critical mass, was surrounded by conventional explosives, when they ignited the resulting shock wave would compress the plutonium. When plutonium was compressed, the atomic particles would be forced into closer contact with each other, and fission would occur. Not only was Teller's design relatively safe; it also used less of the scarce supplies of plutonium.

When Teller was redesigning the hydrogen bomb in the early 1950s, he was confronted with the problem of getting the hydrogen isotopes to undergo a fusion reaction. In nature, fusion reactions are triggered in the sun and stars by extremely high temperatures, but nothing available, including an atomic bomb, could create a high enough temperature on the earth.

Teller realized that compressing the hydrogen would allow a fusion reaction to be triggered at a lower (although still very high) temperature. Fission reactions created massive amounts of x-rays, which generated both pressure and heat. In his design, he placed a hydrogen core inside a hollow, sphere-shaped atomic bomb covered with a heavy metal case that would, at least for a short time, prevent the x-rays from escaping. When the atomic bomb was triggered by conventional explosives, the x-ray radiation released was intense enough to compress the hydrogen. At the same time, the fission reaction generated a tremendous amount of heat. The heat and the pressure forced the hydrogen to undergo fusion.

Some of the scientists who were responsible for building the atom bomb. Teller is standing on the far left; his colleague Enrico Fermi is seated, second from right.

In 1943 Teller moved to a top-secret research facility built in Los Alamos, New Mexico, to house the scientists working on the atomic bomb. Teller, who was accustomed to the freedom of universities, did not adjust well to the hierarchical environment of Los Alamos. He also disagreed with choice of focus there. He believed that the primary focus of the scientists' work should be the hydrogen bomb because it would be more powerful. Indeed, some work was done on the hydrogen bomb at Los Alamos. However, most of the scientists focused on the atomic bomb because it was considered more likely to become a workable device.

Despite the disagreements, Teller did important work on the implosion trigger that was used to set off atomic bombs made of plutonium. He also studied how the escape of x-ray radiation would affect the power of the bomb; these discoveries would be important later in developing a hydrogen bomb.

By August 1945, two atomic bombs had been built and were ready to be used. Germany had already surrendered, but Japan had not. On August 6, 1945, the United States dropped an atomic bomb on Hiroshima, Japan; on August 9, the United States dropped an atomic bomb with an implosion trigger on Nagasaki, Japan. Both bombs were extremely destructive, causing more than two hundred thousand deaths. On August 14, Japan surrendered.

After the war, most of the physicists at Los Alamos were happy to stop working on bombs and return to their peacetime work. Teller, however, was sure that the communist government of what was then the Soviet Union posed a serious threat to the United States, and he thought the U.S. government should pursue the development of the hydrogen bomb. He was in the minority, however, and research into the hydrogen bomb stopped once the war ended.

THE RACE FOR THE H-BOMB

Teller began teaching at the University of Chicago, but in 1949, very early in the Cold War between the United States and the Soviet Union, he returned to Los Alamos, where scientists were working on ways to make the atomic bomb smaller and safer. In the fall of that year, U.S. monitoring equipment detected evidence that the Soviet Union had detonated an atomic bomb.

Teller believed that the Soviet Union would need little time to develop a hydrogen bomb, and that the United States had to develop one first. Many other physicists, however, thought that the United States should refrain from building a hydrogen bomb, to avoid goading the Soviet Union into building one of its own.

The first test of Teller's hydrogen bomb produced this mushroom cloud over Elugelab Island in 1952; the explosion vaporized the island.

An influential government advisory committee of scientists came out against the hydrogen bomb, so Teller, who was becoming politically influential, went directly to various elected officials to garner support. He ultimately succeeded, and in January 1950, President Harry Truman announced a program at Los Alamos to develop the hydrogen bomb.

During World War II, the scientists at Los Alamos had drawn up a design for a hydrogen bomb that relied on a fission reaction to trigger fusion. When researchers examined that design more closely in 1950, however, they discovered that it probably would not work and that the fission reaction would not set off a fusion explosion.

In late 1950, Teller completely reworked the hydrogen bomb, producing a design that would enable the fission reaction to cause fusion. In 1952, the United States tested Teller's design on the Pacific island of Elugelab—the resulting explosion wiped out the one-mile (1.6-km) wide island. Teller's concern that the Soviet Union would soon develop a hydrogen bomb proved valid: in 1953, the Soviet Union announced that it, too, had exploded a hydrogen bomb.

AFTER ELUGELAB

By the time of the test at Elugelab , Teller had quit Los Alamos because he felt the director was not moving aggressively enough to develop the hydrogen bomb. He imagined that a second government nuclear-weapons laboratory would force Los Alamos to be more productive, and he soon found allies at the University of California. In the summer of 1952, Teller moved to Livermore, California, where he became one of the founders of what is now the Lawrence Livermore National Laboratory.

From the 1950s onward, Teller became an increasingly public figure. His opinions were controversial: he was ardently opposed to nuclear test bans and treaties to limit nuclear weapons (he believed that the Soviet Union would secretly circumvent such agreements), and he

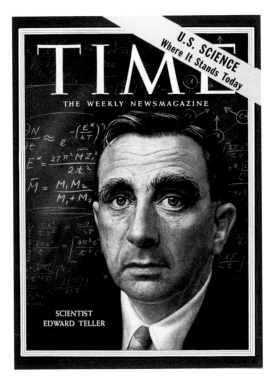

Teller's support of nuclear weapons made him a divisive public figure; he was featured on the cover of Time *magazine in 1957.*

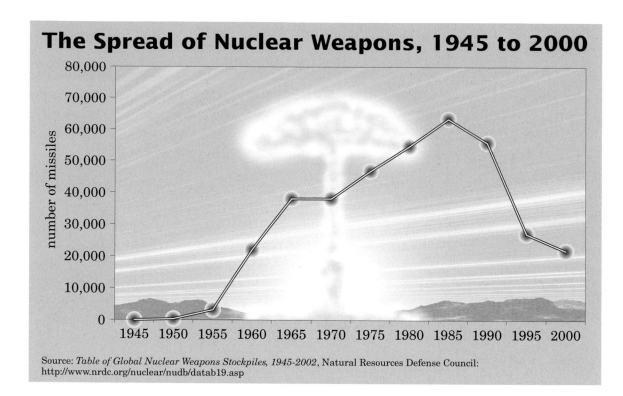

The Spread of Nuclear Weapons, 1945 to 2000

number of missiles (y-axis: 0 to 80,000)

x-axis: 1945 1950 1955 1960 1965 1970 1975 1980 1985 1990 1995 2000

Source: *Table of Global Nuclear Weapons Stockpiles, 1945-2002*, Natural Resources Defense Council: http://www.nrdc.org/nuclear/nudb/datab19.asp

Nuclear weapons stockpiles grew steadily throughout the Cold War era, peaking in the 1980s but then falling throughout the 1990s.

was equally outspoken in his support for nuclear energy and the use of nuclear explosions for peaceful purposes. When the movie, *Dr. Strangelove or: How I Learned to Stop Worrying and Love the Bomb*—a satire about nuclear war—came out in 1964, the rumor arose that Teller was the model for the irresponsible character Dr. Strangelove.

Teller had many opponents, but he also had many supporters—among them Ronald Reagan. After Reagan was elected president in 1980, he initiated the Strategic Defense Initiative (called Star Wars by its detractors), a program that Teller had been espousing for 20 years to develop defenses against attacks by nuclear missiles. In 1983, Teller received the National Medal of Science, and in July 2003, he received

TIME LINE

1908	1926	1930	1935	1939
Edward Teller born in Budapest, Hungary.	Teller attends school in Karlsruhe, Germany.	Teller receives his doctorate from the University of Leipzig.	Teller becomes a professor at George Washington University.	German scientists split the atom.

the Presidential Medal of Freedom, the nation's highest civilian honor. In September of that year he died at the age of 95 in Stanford, California.

Teller firmly believed that nuclear weapons would keep the United States safe. During the Cold War between the United States and the Soviet Union, the fact that each side could completely obliterate the other in a nuclear war was believed to prevent both from using nuclear weapons in a conflict. Some historians argue that the arms race (the constant pressure on the Soviet Union and the United States to keep up with each other militarily) forced the Soviet Union to bankrupt itself, hastening its own dissolution in 1991.

In the early 1990s, the Cold War ended, yet more than twenty thousand nuclear warheads are believed to remain in existence, raising concerns that a terrorist organization could steal them. Over the years more countries have developed nuclear weapons, and the technology is likely to continue to spread, perhaps to countries with irresponsible or unstable leadership.

—Mary Sisson

Further Reading

Books

Blumberg, Stanley A., and Gwinn Owens. *Energy and Conflict: The Life and Times of Edward Teller*. New York: Putnam, 1976.

———, and Louis G. Panos. *Edward Teller: Giant of the Golden Age of Physics*. New York: Scribner's, 1990.

Goodchild, Peter. *Edward Teller: The Real Dr. Strangelove*. Cambridge, MA: Harvard University Press, 2004.

Teller, Edward. *Better a Shield Than a Sword: Perspectives on Defense and Technology*. New York: Free Press, 1987.

TIME LINE

1941	1949	1952	1983	2003
Teller becomes a U.S. citizen and contributes to the development of the atomic bomb.	Teller begins work on the hydrogen bomb.	The first hydrogen bomb is tested over Elugelab Island.	Teller receives the National Medal of Science.	Teller dies.

————. *Memoirs: A Twentieth-Century Journey in Science and Politics.* Cambridge, MA: Perseus, 2001.

Web sites

A History of National Security
    A historical exhibit from the Los Alamos National Laboratory.
    http://www.lanl.gov/history/index.shtml
Lawrence Livermore National Laboratory: About LLNL
    Historical and contemporary information about the defense lab Teller helped found.
    http://www.llnl.gov/llnl/about/
Race for the Superbomb
    An exhibit on the development of the hydrogen bomb by PBS's *American Experience.*
    http://www.pbs.org/wgbh/amex/bomb/index.html

See also: Fermi, Enrico; Military and Weaponry; Nobel, Alfred.

NIKOLA TESLA

Developer of electric power

1856–1943

The farsighted work of Croatian-born inventor Nikola Tesla failed to win proper recognition during his lifetime. A visionary who developed the modern electrical power system, Tesla also pioneered wireless (radio) communication. Tesla was a prolific inventor who received hundreds of patents in at least two dozen countries, inspiring such developments as fluorescent lamps, robotics, radar, and space weapons. Yet others gained the credit for nearly everything he invented.

EARLY YEARS

Nikola Tesla was born in Smiljan, Croatia, on July 9, 1856. His father was a priest with a huge library, where Nikola locked himself away to learn about the world. Tesla described his mother as "an inventor of the first order": she made numerous tools and machines to help with the house-work, including a mechanical eggbeater. Tesla quickly distinguished himself as a brilliant student: he had a photographic memory and could do calculus, a complex form of math, in his head. One of his first inventions was a machine powered by june bugs, a type of beetle.

Tesla's father wanted him to be a priest, but Tesla was determined to become an engineer. After attending high school in the town of Karlstadt (Karlovac), he studied electrical engineering at a polytechnic (a school devoted to applied sciences and technical skills) in Graz, Austria. In Graz, he first began to think in detail about electric motors and generators; motors used the energy in electricity and magnetism to make a wheel spin, whereas generators worked in the opposite way, making electricity from movement. After finishing at Graz, Tesla entered the University of Prague, in what is now the Czech Republic, and graduated in 1880.

A WALK IN THE PARK

Electricity was then still a new technology. No homes were electrically powered and people encountered electricity only through communications apparatus, including telegraph and telephone lines that used electric currents to carry messages down miles of cable. Tesla's first job involved helping to set up the telephone network in Budapest, Hungary.

One afternoon in 1882, Tesla was walking in a Budapest park when he suddenly had an idea for a new electric motor, which he immediately drew with a stick in some sand. As he later recalled: "The idea came like a flash of lightning and in an instant the truth was revealed." In a normal motor, the electric current always flows in the same direction, like a river; this type of current is called direct current (DC). Tesla realized that if the current could be made to reverse

A photograph of Nikola Tesla, originally published in Illustrated London News in 1900.

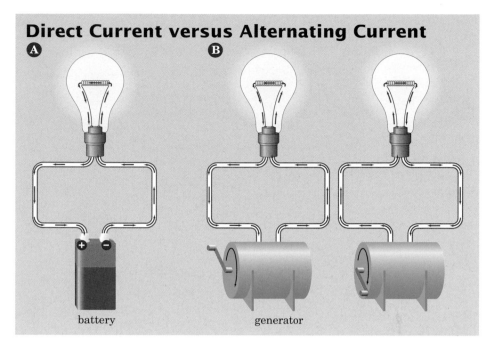

Direct Current versus Alternating Current

A

B

battery   generator

Direct current (DC), which generates power in most batteries, flows in one direction (A). Alternating current (AC), which generates power in bigger devices such as generators used for power plants, reverses its flow after a specific period of time (B).

direction, flowing alternately one way and then the other, like the ocean tides, the motor would work more effectively. This continually reversing type of electricity, which Tesla pioneered, was known as alternating current (AC).

WORKING WITH EDISON

For the next two years, Tesla moved around Europe. In 1882, he worked for Continental Edison in Paris, France. The following year, he went to work for the company in the city of Strasbourg and in his spare time built the first electric motor to use alternating current. The company's founder, Thomas Edison (1847–1931), was then pioneering large-scale electric power in the United States. Edison's development of practical electric light, in the 1870s, had created a great demand for electricity. In 1882, Edison opened the world's first commercial electric power plant on Pearl Street in New York City using DC.

Tesla was determined to prove that AC was better and immigrated to the United States (in his words, the "land of golden promise") in 1884 to work with Edison. Reputedly, he arrived in New York City with four cents in his pocket, some poems he had written, a sketch for a flying machine, and a letter of introduction to his hero. Edison quickly set Tesla to work, promising him $50,000 if he could make a better power-production system. Tesla immersed himself in the task, working from 10:30 a.m. each day straight through until 5:00 a.m. the next morning. Edison, himself a workaholic, was impressed: "I have had many hard-

working assistants, but you take the cake." Yet when Tesla delivered an improved power system, just a few months later, Edison said that the $50,000 offer he had promised was a practical joke. Tesla, who was shocked and disappointed, immediately resigned.

Tesla's resignation turned out to be very costly for Edison, however, because the following year Tesla sold the rights to his AC power system to George Westinghouse (1846–1914). After a lengthy battle between Edison and Westinghouse and their rival power systems, Westinghouse won the contract to build a huge power plant at Niagara Falls in 1893. Edison's DC system was effectively sidelined, and most of the world's power has been generated and supplied using AC ever since. Even so, Edison, not Westinghouse or Tesla, is honored as the inventor of electric power.

WIRELESS WORLD

Using the money from his deal with Westinghouse, in 1889 Tesla established the first of several laboratories in New York City. He continued his experiments in alternating current. Alternating current worked by reversing the electricity flow many times per second, at a particular frequency. Tesla became interested in exploring what happened when he made the frequency much greater.

In 1891, Tesla invented what became known as the Tesla coil: an electrical transformer that could increase the frequency, or voltage, of AC electricity hundreds of thousands of times. Operating at such high frequencies, the coil could generate electromagnetic radiation, a type of energy, made from waves of electricity and magnetism, now known to carry such radiation as light, radio waves, and x-rays. Tesla used his coil to power some of the first neon and fluorescent lights and also took "shadowgraph" photographs with it, similar to the x-ray pictures later invented by German physicist Wilhelm Röntgen (1845–1923).

Radio (also known as wireless) was another use for the new coil. Tesla carried out his first radio experiments in 1892, several years before Guglielmo Marconi (1874–1937), the man usually credited with the invention. By 1895, Tesla had developed a device that could send radio signals 50 miles (80 km), though a fire at one of his laboratories slowed his progress. Consequently, Marconi filed the first radio patent in Britain in 1896; Tesla filed the following year in the United States. In 1897, Tesla gave the first demonstration of wireless remote control at an electrical exhibition in New York City. Using radio signals, he baffled onlookers by making a model boat flash its lights and move around a small indoor pond apparently all by itself. This was an early example of robotics, more than 20 years before the term *robot* was coined.

After Tesla's U.S. patent was granted in 1900, Marconi could no longer file for American patents. For a time, Tesla seemed about to make history as the pioneer of American radio. With a $150,000 investment from banker J. P. Morgan, he constructed a broadcasting tower on Long Island and planned to set up a worldwide radio station there. However, the project fell through when Morgan withdrew his backing. In 1904, the U.S. Patent Office inexplicably overturned Tesla's radio patents, awarding them to Marconi instead. Two years later, electrical engineer Lee De Forest (1873–1961) invented the Audion or triode (three-legged) vacuum tube, a tiny "electronic ear" that could detect radio waves passing through the air and with other circuitry could turn them back into sounds. De Forest's invention made possible the manufacture of inexpensive radio receivers and earned him, not Tesla, the title "father of radio."

A *Tesla induction motor; the motor uses AC current to produce a rotating magnetic field, which makes the motor turn.*

MORE INVENTIONS

By the late 1890s, and still only in his mid-forties, Tesla had helped to pioneer two of the greatest inventions of the 19th century. The next step was to combine his innovations into a wireless power system that could carry electricity to lights or other appliances without cables. During 1899, Tesla tested this idea with remarkable experiments at Colorado Springs (see box, Lightning Man).

In 1912, Tesla outlined plans for a revolutionary new steam turbine: an engine that could turn steam into electricity. Although he intended to compete with the turbines that Westinghouse and Edison had developed, he never perfected the design and soon abandoned it. During

Lightning Man

Tesla carried out his most dramatic experiments in 1899, while trying to prove that he could transmit electric power without wires. He believed that earth's atmosphere could conduct electricity and decided to test this idea by creating enormous amounts of electrical energy at high altitudes. With financial help from El Paso Power Company, he moved to Colorado Springs, one of the highest regions in the United States. There, in the middle of the prairie, he built a giant Tesla coil with an 80-foot (24-m) wooden tower above it and a 142-foot (43-m) metal antenna on top of that, all topped by a large copper ball.

After careful tests, Tesla ordered his assistant to throw the mighty power switch. As the coil flashed blue, it generated 100 million volts of electric power, hurling artificial bolts of lightning more than 135 feet (41 m) long from the metal ball. In experiments like this, Tesla managed to light 200 fluorescent lamps planted in the ground a distance of up to 25 miles (40 km) from the coil without any wires. However, many scientists think that the electricity traveled across the surface of the ground rather than through the air, as Tesla supposed.

Although far from successful, the Colorado experiments were certainly mysterious and controversial. On one occasion, Tesla blew up the local power generator, throwing the whole of Colorado Springs into darkness. Another time, he claimed to have received messages from outer space.

TIME LINE

1856	1880	1882	1884	1889
Nikola Tesla born in Smiljan, Croatia.	Tesla graduates from the University of Prague.	Tesla creates a motor using alternating current.	Tesla travels to New York City to work for Thomas Edison.	Tesla establishes a laboratory.

World War I, he proposed a way of using radio to detect approaching enemy ships, though this was another idea he never managed to develop. Twenty years later, British military engineer Robert Watson-Watt (1892–1973) produced an almost identical invention named *radar*, which revolutionized air and sea warfare.

LATER YEARS

None of Tesla's later inventions were as successful as AC power. His last patent, filed in 1928, titled "Apparatus for Aerial Transportation," was a combination of an airplane and a helicopter. This was more than a decade before Igor Sikorsky (1889–1972) made the first practical helicopter and decades before engineers invented tilt-rotor airplanes. These airplanes had propellers that could point either forward or upward and could take off like helicopters.

Tesla unveiled his last great idea on July 11, 1934, when the *New York Times* ran the front-page headline: "Tesla, at 78, Bares New Death Beam." In his most audacious attempt to show that electricity could travel through air, he proposed an almighty electric gun that could destroy 10,000 airplanes at a range of more than 200 miles (322 km). Although never developed in Tesla's lifetime, energy weapons, such as laser beams mounted in space, are thought by many to represent the future of warfare.

As Tesla grew older, his inventions became increasingly eccentric—including a machine for time travel and a camera that could photograph thoughts. He befriended pigeons and became fascinated with the number three. He was afraid of round objects and also of germs, refusing to shake hands with people for fear of contamination. Always hardworking and a man of extraordinary energy, he came to believe that neither sleep nor food was really necessary. Instead, he thought his body and mind could get their nourishment from a vibrating electric plate. His

Tesla's image appears on the 100-dinar banknote, issued by the National Bank of Serbia. Tesla was also pictured on currency in the former Yugoslavia.

TIME LINE (continued)

1891	1897	1899	1928	1943
Tesla invents the Tesla coil.	Tesla demonstrates wireless remote control.	Tesla tests a wireless power system.	Tesla patents a flying machine.	Tesla dies.

neglected health led to his death on January 7, 1943, at the age of 86. When Tesla died, he was virtually penniless and many of his inventions were unknown. Few people who used electric power or radio remembered the brilliant visionary who had played such an important part in these inventions.

However, Tesla was not without admirers. Three days after his death, the mayor of New York City delivered a eulogy on the radio, describing Tesla as "a distinguished scientific genius and a poet of science." Two days later, Tesla's funeral was attended by 2,000 people, including many distinguished scientists, inventors, and other notables. Appreciation of Tesla's work has increased with time: the United Nations Educational, Scientific, and Cultural Organization (UNESCO) declared the year 2006, the 150th anniversary of Tesla's birth, to be the Year of Tesla. Also in 2006, the largest airport in Serbia was renamed the Belgrade Nikola Tesla Airport in his honor.

—Chris Woodford

Further Reading

Aldrich, Lisa. *Nikola Tesla and the Taming of Electricity.* Greensboro, NC: Morgan Reynolds, 2005.

Cheney, Margaret. *Tesla: Man out of Time.* New York: Touchstone, 2001.

Seifer, Marc. *Wizard: The Life and Times of Nikola Tesla.* New York: Kensington/Citadel, 2001.

See also: Edison, Thomas; Energy and Power; Faraday, Michael; Morse, Samuel; Watson-Watt, Robert.

LINUS TORVALDS

Inventor of the Linux
operating system

1969–

The growth of the Internet since the mid-1990s has enabled people to work together in many new ways, even when they live on opposite sides of the world. Thousands of inventive computer users have used the Net to contribute to Linux (pronounced Linn-ucks), a free rival to the popular Microsoft Windows operating system. Linux is named for Linus Torvalds, the Finnish computer programmer who created it in 1991.

EARLY YEARS

Linus Benedict Torvalds was born in Helsinki, Finland, on December 28, 1969, and was named for Linus Pauling (1901–1994), an American chemist and Nobel Prize winner. His father, Nils, works in broadcasting; his mother, Mikke, is a journalist. In his 2001 autobiography, *Just for Fun*, Torvalds describes himself as having been "an ugly child," "a beaverish runt," and "a nerd—before being a nerd was considered to be a good thing."

During the early 1980s, he became interested in computers and acquired a Commodore VIC-20, an inexpensive and quite basic home computer largely designed for playing games. Later, he bought a more powerful machine and started to tinker with its operating system. The operating system is the basic software that controls a computer's hardware and its applications. Torvalds had no idea that his growing interest in operating systems would dramatically change not only his life but also the world of computing.

INVENTING LINUX

In 1988, Torvalds began studying at the University of Helsinki for a degree in computer science that he took an extraordinary eight years to complete. He began normally enough, doing much of his college work using the UNIX operating system, which had become popular in universities in the 1970s. Then, in 1990, he bought an IBM personal computer (PC) that ran DOS (disk operating system), a different operating system that had been developed by Microsoft. Torvalds thought DOS was greatly inferior to UNIX and wondered if UNIX could be made to work on the IBM PC instead.

He was not the first to try to adapt UNIX for the PC. An American academic, Andrew Tanenbaum, had already written Minix, a UNIX-like operating system for the IBM PC.

Linus Torvalds at a computer show in 2000.

After inspecting Minix closely, Torvalds was not satisfied and decided to create his own version of UNIX instead. Throughout the spring and summer of 1991, he shut himself in his bedroom in his mother's apartment and worked on the code, the list of instructions that make up a computer program, around the clock. According to his mother, there was nothing very strange about this behavior: "Just give Linus a spare closet with a good computer in it and feed him some dry pasta, and he'll be perfectly happy."

In August of that year, he posted an announcement on a computer bulletin board: "I'm doing a (free) operating system (just a hobby, won't be big and professional . . .)." After he wrote the kernel, or core, of his operating system, he made it freely available on the Internet, allowing anyone to download it or submit improvements. Torvalds originally called his system Freax, combining three words that seemed to sum up what he was attempting: Free, Freak, and UNIX. A friend later changed the name to Linux, and the new name stuck.

THE RISE OF LINUX

As the 1990s progressed, Linux became an astonishing phenomenon. Tens of thousands of programmers joined the effort to develop it into an operating system that could challenge Windows, the operating system that Microsoft produced after DOS. Like the conductor of some enor-

The desktop of the Xandros operating system; Xandros is a type (or distribution) of Linux that is popular with many businesses because of its visual similarity to Microsoft Windows.

mous worldwide orchestra, Linus Torvalds took time off from working on his college degree to direct the effort. Some of the programmers helped to build up the Linux kernel; others developed or modified application programs to make them compatible with Linux. However, Torvalds always remained in charge in a friendly but determined way. As he said to one programmer about a proposed change to Linux in 1996: "If you still don't like it, that's OK: that's why I'm boss. I simply know better than you do."

The success of Linux earned Torvalds widespread admiration in the computing world. Under his guidance, Linux had become not just an affordable operating system but one that had other advantages, too. It could be quicker and more powerful than Windows and more robust, less likely to "crash" when something went wrong. It was also more secure against viruses and similar threats. Advantages like these made Linux increasingly attractive to computer users.

Inventing Together

Linus Torvalds has helped to popularize a whole new approach to inventing. Several aspects make his Linux program different from an ordinary invention. Most obviously, it is a free product—free in the sense that it is usually given away. This is possible because the programmers who work on Linux want to be part of a community effort that helps people; they do not especially want to make money. Unlike most inventions, Linux is free from copyright and patent restrictions; thus it can spread and evolve much more quickly than an invention bound by legalities.

Linux is also different in being the product of thousands of inventive minds. Although Linus Torvalds started this effort alone, an estimated that 98 percent of the Linux code has since been written by others. Because so many people are involved in the project, Linux has developed much more quickly than would have been possible had Torvalds worked alone. This kind of "group inventing" is possible largely because of the Internet, which allows people to chat, e-mail, and exchange their creations almost instantly. The huge success of Linux has inspired many others to try this method of creation; the Wikipedia project, for example, is a free Internet encyclopedia comprising several million articles written by tens of thousands of enthusiasts.

By January 2000, Linux was such a trustworthy piece of software that IBM, the world's biggest computer company, announced that it was going to make all its computers run on the system. Since then, Wal-Mart has started selling PCs that run Linux, and many cell phones and palmtop computers also have switched to the system. Linux has become a truly global phenomenon: many of the world's governments are opting to use it, rather than being tied to American-engineered Microsoft products. Linux is even being used in low-price, wind-up laptop computers being manufactured for children in developing countries.

TROUBLE AHEAD?

In 1991, Linux had just one user, Linus Torvalds; a decade later, millions had switched to the system—largely because it was free. Linux is free in two senses: it is available for free download from the Internet and it is also free from copyright, so people can share and modify it. Programs like this are known as "open source": the original software, or source code, is freely available to others to use as they wish.

Open-source programs can be very attractive to users. However, big companies that sell software for a fee, such as Microsoft, have sensed a growing threat from the huge army of volunteer programmers who are developing software this way. One such company, Santa Cruz Operation (SCO), has even tried to sue, because it claims that parts of Linux are closely based on UNIX, which it now owns. Because Torvalds and his developers have built their system from huge amounts of donated program code, it is difficult—perhaps impossible—for them to check that none of their code had been copied from software developed by other companies. If corporations such as Microsoft prove that their

Students at Nooitgedacht Middle School in Cape Town, South Africa, using Linux computers in 2004.

TIME LINE

1969	1991	1996	1997	2000	2003
Linus Torvalds born in Helsinki, Finland.	Torvalds completes the kernel of Linux.	Torvalds finishes his computer science degree from the University of Helsinki.	Torvalds begins to work at Transmeta Corporation.	IBM announces that all its computers will run on Linux.	Torvalds begins work at the Open Source Development Labs.

copyrighted software or patents have been wrongly used inside Linux, they could take legal action against Torvalds, Linux developers, or Linux users, possibly with disastrous consequences.

In 1996, Torvalds finally finished his degree at the University of Helsinki. The following year, he announced that he was moving to California to work as a software developer at the Transmeta Corporation. He remained there until June 2003 before transferring to the Open Source Development Labs in Oregon. Various computer firms that sell Linux-based software have donated stock options to him in appreciation of his work. However, he has so far refused lucrative offers of employment that could have made him a very rich man, preferring to focus on improving his free software.

—Chris Woodford

Further Reading

Books

Moody, Glynn. *Rebel Code: Linux and the Open Source Revolution*. New York: Penguin, 2002.

Torvalds, Linus, and David Diamond. *Just for Fun: The Story of an Accidental Revolutionary*. New York: HarperCollins, 2001.

Web site

OpenSources: Voices from the Open Source Revolution
    Interviews with open-source pioneers, including Linus Torvalds.
    http://www.oreilly.com/catalog/opensources/book/toc.html

See also: Berners-Lee, Tim; Communications; Computers; McAfee, John; Omidyar, Pierre.

TRANSPORTATION

Important inventions bring social changes, but arguably none more dramatic than those caused by developments in transportation. Between 3500 BCE and modern times, inventors have taken human society on a journey of innovation from the wheel to the space rocket. People and goods have moved around the world in almost every imaginable way: over land and sea, through the air and space, and now, in the modern age, along the virtual "highways" of the Internet.

THE BEGINNINGS OF TRANSPORTATION

The hectic modern movement of people and goods would seem extraordinary to prehistoric peoples—who mostly traveled on foot, carrying loads in their arms, on their backs, or slung from poles between them. Prehistoric peoples tended to locate their settlements near rivers or coasts; long-distance travel was done mostly by water, using rafts, boats made from skin and bark, and dugout canoes. Around 7000 BCE, the most sophisticated piece of transportation technology was the sled, dragged by humans. American Indian tribes invented the A-shaped travois: a large animal skin tied to two long wooden poles that was fastened to either side of an animal and dragged on the ground behind the animal; however, the era of its invention is uncertain. Archaeologists believe human societies first began using animals to pull loads between 5000 and 3000 BCE.

An Atsina Indian travois, photographed by Edward Curtis in 1908.

Such animals became known as beasts of burden, the "engines" of primitive transportation. The wheel was one of the later developments in prehistoric times, invented around 3500 BCE in the Middle East. By 3000 BCE, wagons and sailing vessels were in common use.

Around 3000 BCE, sailing vessels became increasingly sophisticated, developed by the ancient Egyptians, Minoans, and Phoenicians of the Mediterranean region. The Egyptians were pioneers in ship construction: instead of digging small boats out of trees, they fastened planks onto trees and built upward. The Phoenicians made some of the first sturdy oceangoing ships. The ancient Greeks, who became the dominant civilization in that region around 400 BCE, developed huge, swift ships powered by sails and dozens of oarsmen. Like the Greeks, the Romans (whose empire lasted between 27 BCE and 495 CE) used enormous ships to ferry goods around the territories they had conquered in what is now Europe and the Near East. They also developed their ships into formidable weapons of war.

Land transportation also advanced. In Asia, the Chinese and the Persians began constructing the first primitive dirt-track roads between major settlements between around 1000 and 500 BCE. The Romans constructed the world's first network of highways using gravel and stone. Some of these roads, widened and paved, still exist as modern highways.

SHIPS

Ships became more sophisticated during the Middle Ages in Europe (which lasted from the fall of the Roman Empire to about 1500); they were sturdy and spacious, and moved faster with more oarsmen and larger areas of sail. These advances allowed Christopher Columbus, Vasco da Gama, and Ferdinand Magellan to make ocean voyages during the fifteenth and sixteenth centuries. Navigation also improved. The

In 1992, the 500th anniversary of Columbus's voyage to the Americas, replicas of the Niña, the Pinta, and the Santa Maria sail off the coast of Florida, while the space shuttle Endeavor sits on the launch pad in the background.

Chinese invented the most important navigational device, the magnetic compass, around 1200. In 1537, a Belgian mapmaker, Gerardus Mercator (1512–1594), revolutionized navigation when he developed the Mercator projection: a way of representing the three-dimensional earth on a two-dimensional sheet of paper using lines of latitude and longitude.

A revolution in shipping began in the 18th century when inventors started to explore ways of powering boats using coal-fired steam engines. The first steamboat was launched in 1787 by American inventor John Fitch (1743–1798). It was approximately 45 feet (14 m) long and sailed up the Delaware River at about six miles per hour (10 km/h). Some years later, another American inventor, Robert Fulton (1765–1815), developed a more practical and efficient steamboat. On August 18, 1807, his 150-foot (46-m) steamboat *Clermont* made a 300-mile (483-km) round-trip from New York City to Albany in 62 hours.

These vessels demonstrated that steam was a practical source of power for ships and boats. Just over a decade after Robert Fulton's voyage, in 1819, the steamship *Savannah* crossed the Atlantic from Savannah, Georgia, to Liverpool, England in 29 days. However, other inventions were necessary to make great 19th-century steamships a reality. One was the iron hull, the basic metal framework of a ship. Another important advance, the screw propeller, was developed by American inventor John Ericsson

(1803–1899) and first used in 1844. During the 18th century, daring engineers such as Isambard Kingdom Brunel (1806–1859) built enormous transatlantic steamships, which could cross the Atlantic Ocean in about two weeks. The biggest of them all, Brunel's *Great Eastern*, completed in 1858, was almost 700 feet (213 m) long, had smokestacks 100 feet (30 m) high, and carried four thousand passengers. Iron ships later gave way to ships made from steel. This stronger and more versatile material was developed in the mid-19th century by English inventor Henry Bessemer (1813–1898).

RAILROADS

Most people's first experience with steam energy came not from ships but from

The locomotive Jupiter moving along the tracks at the Golden Spike National Historic Site in Utah, where eastbound and westbound railroads were united to form the transcontinental railroad in 1869.

The Cost of Transportation

Transportation lies at the heart of modern society and is equally crucial to national economies. Diesel and jet engines power the trucks, locomotives, ships, and planes that carry goods across oceans and continents by land, sea, and air. Inventions such as these, combined with improvements in communications technology, have made it possible for companies to operate effectively in more than one country. During the 20th century, international transportation became so inexpensive, rapid, and well-organized that it was often cheaper for multinational companies to manufacture goods abroad and import them than to manufacture them at home. Similarly, inexpensive transportation has opened up new export markets. Globalization, as these and similar trends have become known, has led to dramatic social changes. Many manufacturing industries have relocated to Asian countries such as China.

Greater use of transportation has come at a price, however. Environmental problems such as smog pollution and global warming are related to the burning of fossil fuels in many millions of car, truck, ship, and airplane engines. Much of the world's oil is supplied by countries of the Middle East, a politically unstable region of the world. So many vehicles are now on the move that traffic congestion has itself become a major environmental and economic problem.

Although globalization has brought many companies greater wealth and many individuals a greater choice of affordable goods, some people believe it has undermined jobs in western countries and encouraged environmental pollution and sweatshops in developing nations. The transportation of goods and people has grown enormously since the 20th century, but that growth cannot continue to be sustained by the earth. Some people believe the solution to these problems may be to reduce the need for travel by using advanced communications technologies such as the Internet. Others think environmentally friendly cars, trucks, ships, and planes may address the problem. One thing is certain: inventors are sure to discover better forms of transportation in the future.

steam engines on land. The first practical steam engine was developed in 1705 by English engineer Thomas Newcomen (1663–1729) and was originally intended for pumping wastewater out of mines. Within a few decades, other inventors had tried using steam engines to drive vehicles, but the technology was not advanced enough for these attempts to be successful. Around 1770, a French army engineer, Nicolas Cugnot (1725–1804), invented the world's first self-propelled road vehicle: a steam-driven automobile-tractor. Although it moved slowly (top speed was two miles per hour, or 3.2 km/h), it was not readily maneuverable—when it crashed into a wall and overturned, it was consigned to history.

A breakthrough came in the late 18th century when English engineer Richard Trevithick (1771–1833) developed more compact high-pressure steam engines. At the dawn of the 19th century, on Christmas Eve of 1801, Trevithick completed the first steam-powered vehicle that could carry passengers; three years later, he built a crude steam locomotive. Twenty more years passed before another English engineer, George Stephenson (1781–1848), made the first practical railroad locomotive. In 1825, the Stockton to Darlington Railway near Newcastle, England, began the world's

first steam railroad using Stephenson's engine. The design of steam locomotives improved gradually through the following decades, with innovations being made by many different engineers. In 1872 Canadian-born engineer Elijah McCoy (1843?–1929) invented a way of lubricating (oiling) the moving parts in steam engines automatically, thus permitting them to run faster and more reliably. Granville Woods (1856–1910) was a developer of electric railroads.

The railroads spread rapidly during the 19th century not just because of better locomotives. Ingenious feats of engineering and thousands of workers helped to lay many miles of track over plains; through, around, and over mountain ranges; and across deep valleys. In the United States, one pioneer of this work was German-born American engineer John Roebling (1806–1869), inventor of the suspension bridge. In England, during the 1830s, Isambard Kingdom Brunel made his reputation as the engineer of Britain's Great Western Railway: a railroad that ran hundreds of miles from London, the English capital, to Cornwall, on the southwest coast. The Baltimore and Ohio Railroad became the first railroad in the United States when construction began in 1828. During the second half of the 19th century, railroad

TIME LINE

5000–3000 BCE	ca. 3500 BCE	3000 BCE	1000 BCE	1200 CE	1537
Humans begin using animals to pull loads.	The wheel is invented.	Wagons and boats are in use.	Dirt roads are constructed in Asia.	Magnetic compass is invented.	Gerardus Mercator develops his Mercator map projection.

tracks were laid across the entire North American continent. The first transcontinental railroad was completed in Utah on May 10, 1869. By 1930, when the railroad network reached its peak, 430,000 miles (692,018 km) of track had been laid, enough to stretch from New York City to Los Angeles 175 times.

ROADS

Railroads revolutionized long-distance transportation and played an important part in opening up western states to settlers. However, trains could visit only those places where there was track and, despite the efforts of engineers such as Cugnot and Trevithick, steam-driven road vehicles never became popular. For shorter distances and more local journeys, the horse remained the most important form of transportation until the end of the 19th century.

Human power offered one solution to the problem of local travel in the form of the bicycle. An early, primitive bicycle, called the drasine or the dandyhorse, was patented by German tax collector Karl von Drais (1785–1851) in 1818. Instead of pedaling, the cyclist simply straddled the frame and ran along. A better design, credited to French inven-

tor Pierre Michaux (1813–1883) and others, appeared in the 1860s. Michaux's "boneshaker" was aptly named; its iron and wooden wheels had no suspension and offered a most uncomfortable ride, but it was propelled by pedaling and was thus faster than the dandyhorse. As bicycles became popular, they became faster and more comfortable. Rubber tires first appeared around the 1870s, greatly increasing demand for a tougher kind of "vulcanized" rubber that had been developed in 1839 by American inventor Charles Goodyear (1800–1860).

Bicycles were versatile and affordable, but how far and how fast riders could pedal were limited, as was the amount and weight of goods a bicycle could carry. A more compact source of power than the steam engine was needed. Steam engines worked by an external combustion process. An open-air fire burned coal, which heated water in a boiler to make steam. The expanding steam pushed a piston back and forth in a large metal cylinder (like a bicycle pump moving in and out) to produce power and motion. In the seventeenth century, a Dutch physicist, Christian Huygens (1629–1695), realized that burning the fuel inside the cylinder would be more efficient and came up with the idea of internal combustion. Another two centuries were to pass, however,

TIME LINE (continued)					
1705	**1770**	**1783**	**1787**	**1801**	**1807**
Thomas Newcomen builds a steam engine.	Nicolas Cugnot builds the first self-propelled vehicle.	The Montgolfier brothers fly the first hot-air balloons.	First steamboat created by John Fitch.	Richard Trevithick builds a steam-powered passenger vehicle.	Robert Fulton's steamboat *Clermont* makes a 300-mile (483-km) voyage.

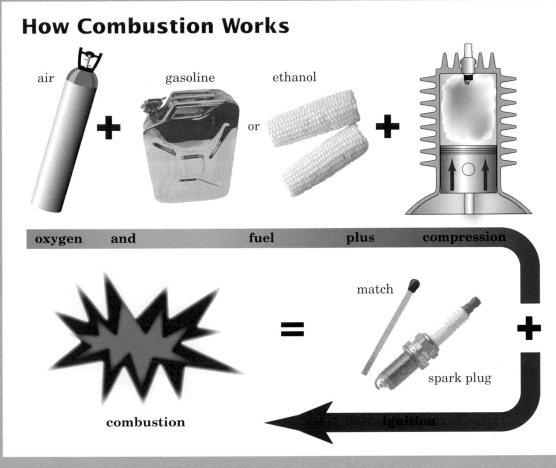

How Combustion Works

air **+** gasoline ethanol **or** **+**

oxygen **and** fuel **plus** compression

match

spark plug

= **+**

combustion **ignition**

The process of combustion is fundamental to motorized travel.

before this idea would become one of the most important inventions of all time: the gasoline engine.

Numerous inventors played a part in the development of the gasoline engine, beginning in 1860 with French engineer Joseph Étienne Lenoir (1822–1900). Seven years later, German engineer Nicolaus August Otto (1832–1891) developed an efficient gasoline engine. It was the first real

TIME LINE

1818	1825	1830s	1838	1844	1858
The drasine is invented by Karl von Drais.	The Stockton to Darlington railway is opened.	Isambard Kingdom Brunel designs the Great Western Railway.	Charles Goodyear develops vulcanized rubber.	John Ericsson invents the screw propeller.	The steamship *Great Eastern* is completed.

An advertisement from Sweden for the 1936 Ford coupe.

mechanical alternative to inefficient steam engines. In the last few years of the 19th century, another German engineer, Rudolf Diesel (1858–1913), developed a simple, powerful engine that could run on almost any fuel. More robust than the gasoline engine, the diesel engine was soon used in factories, ships, power plants, trucks, submarines, and automobiles.

The smaller gasoline engine led to more compact, personal forms of transportation in the second half of the 19th century. One of the first to explore this area was German engineer Karl Benz (1844–1929), who built his own gasoline engine in 1878. By 1885, he had fastened this machine to a three-wheeled bicycle carriage and produced the Tri-Car—which many people regard as the world's first gasoline-powered automobile. His rival, Gottlieb Daimler (1834–1900), then produced a four-wheeled car. Together, Daimler and Benz ushered in the age of modern motoring. American industrialist Henry Ford (1863–1947) introduced his Model T in 1908. Significantly, Ford developed the assembly line for his factories; this form of manufacturing allowed him to make cars in greater quantity with

less expense; thus he could charge less for his automobiles and more people could afford them. Japanese industrialist Soichiro Honda (1906–1991) performed a similar feat with the motorcycle.

The first cars were basic: Benz's Tri-Car had no roof to protect against the rain, and the engine in Ford's Model T had to be started by a hand crank. Later inventors turned their ingenuity to making cars more automated and refined. Charles Kettering

TIME LINE (continued)

1860s	1867	1869	1872	1885	1900
Pierre Michaux popularizes the bicycle.	Nicolaus August Otto develops a gasoline engine.	The transcontinental railroad is completed.	Elijah McCoy patents a lubricating cup for engines.	Karl Benz constructs the Tri-Car.	Ferdinand von Zeppelin pilots his first dirigible.

(1876–1958) developed the electric ignition (a car's automated starting system, which replaced the hand crank), and Mary Anderson (1855–1940) invented windshield wipers. Others made driving cars safer, including Garrett Morgan (1877–1963), who developed traffic signals; and Allen Breed (1927–1999), inventor of the air bag.

AIR AND SPACE

Archaeologists believe ancient civilizations experimented with flight, but not until 1783 were human dreams of flying realized. That year, French brothers Joseph-Michel Montgolfier and Jacques-Étienne Montgolfier (1740–1810; 1745–1799) flew the first hot-air balloons. Simple spherical balloons had no means of control and blew wherever the wind took them, so they were more of a novelty than a practical form of transportation. By 1900, a German inventor, Ferdinand von Zeppelin (1838–1917), had developed a better design: a lozenge-shaped balloon, built around a metal framework, powered by engines with propellers. This conveyance could be steered and directed; zeppelins were known as airships and could fly for days at a time, but they were soon superseded by a better form of air travel.

Most innovators in transportation developed their revolutionary ideas by studying earlier inventions and determining how to improve them. Karl Benz used a bicycle carriage to form his first car, and Henry Ford was inspired by a steam tractor he saw on his family's farm. Two of the greatest aerospace innovators were similarly inspired: Orville Wright (1871–1948) and Wilbur Wright (1867–1912) began as bicycle makers in a machine shop in Dayton, Ohio, and took their ideas from gliders (unpowered aircraft). In December 1903, they too surpassed the inventions that had fired their imaginations when they became the first to fly an engine-powered airplane. The Wrights, in turn, prompted other inventors to devise better airplanes. One, a Russian-born engineer named Igor Sikorsky (1889–1972), met Wilbur Wright in 1908 and then decided to make an aircraft that could fly to places airplanes could not. He finally realized his ambition in 1943, during World War II, when the world's first practical helicopter went into production.

The modern airplane jet engine was born just before World War II as a result of a rivalry between Englishman Frank Whittle (1907–1996) and German Hans Pabst von Ohain (1911–1998): Whittle developed the first engine, but Ohain made the first flight. Once the potential of

TIME LINE

1903	1908	1926	1943	1969
Orville and Wilbur Wright fly an engine-powered airplane.	The Model T car goes on sale.	Robert H. Goddard launches a rocket.	Igor Sikorsky's helicopter goes into commercial production.	Humans land on the moon.

jet airplanes was realized, engineers were soon making them fly farther and faster. Jet planes became rocket planes, and during the 1950s humans began the conquest of space. Space travel seems a natural evolution from air travel, yet the invention of space rockets predated jet engines by more than a decade. The first space rocket was actually launched in 1926, from a cabbage patch on a farm in Auburn, Massachusetts, by American physicist Robert H. Goddard (1882–1945). Goddard did not live to see astronauts blasting into space or reaching the moon in 1969, but these achievements would not have been possible without him.

In prehistoric times, travel was minimal; in the 21st century, people think nothing of taking jet planes from one side of the world to the other. Inventors will be just as important for tomorrow's transportation. For example, experimental solar-powered cars are already being worked on. Instead of engines, these cars have electric motors; instead of gas tanks, they have hoods, trunks, and roofs covered with solar panels that generate electricity. Such vehicles produce no pollution and cost almost nothing to run. Other inventions currently in development include ship engines that move water using magnetism, and silent supersonic (faster-than-sound) airplanes that use less fuel. Every new form of trans-

portation has affected not just how people travel in the world but how they see and experience it.

—Chris Woodford

Further Reading

Herbst, Judith. *The History of Transportation*. Breckenridge, CO: Twenty-First Century, 2005.

Sutton, Richard. *Eyewitness: Car*. New York: Dorling Kindersley, 2005.

Williams, Harriet. *Road and Rail Transportation*. New York: Facts on File, 2004.

Woodford, Chris. *Ships and Submarines*. New York: Facts on File, 2004.

See also: Anderson, Mary; Benz, Karl; Breed, Allen; Brunel, Isambard Kingdom; Daimler, Gottlieb; Diesel, Rudolf; Ericsson, John; Ford, Henry; Fulton, Robert; Getting, Ivan I.; Harrison, John; Honda, Soichiro; Kettering, Charles; McCoy, Elijah; Mercator, Gerardus; Michaux, Pierre; Montgolfier, Joseph-Michel, and Jacques-Étienne Montgolfier; Morgan, Garrett; Newcomen, Thomas; Otto, Nicolaus August; Ovshinsky, Stanford; Roebling, John A.; Sikorsky, Igor; Stephenson, George; Whittle, Frank; Woods, Granville; Wright, Orville, and Wilbur Wright; Zeppelin, Ferdinand von.

EARL TUPPER

Inventor of Tupperware

1907–1983

Earl Tupper achieved both of his lifelong ambitions—to improve the way people live and to make a fortune as an inventor—when he invented a type of plastic container for storing food. Touted as the greatest advance in food storage since the invention of canning, Tupperware assumed a central place in kitchens throughout America in the 1950s and has since become one of the most recognized household names in the world.

EARLY YEARS

Earl Tupper was born on July 28, 1907, in Berlin, New Hampshire. His father was a semi-successful farmer who also had a passion for inventing gadgets. Nothing much came of his tinkering, although he did receive a patent for a device used to clean chickens. To make ends meet, Tupper's mother, Lulu, operated a boardinghouse in the family's home and also took in laundry from neighbors for additional income.

While Earl Tupper was still a child, the family moved to Shirley, Massachusetts, and opened a greenhouse. From an early age, Earl had an obvious desire to succeed. When he was only 10 years old, he began selling the family's produce door-to-door rather than at a farmer's market because he could sell much more. Upon graduation from high school in nearby Fitchburg in 1925, he continued to work in his family's greenhouse for a couple of years as well as at various odd jobs.

He also began taking correspondence courses (receiving class assignments and submitting results by mail), including one in advertising. Inspired by that course, he urged his family to undertake a more active marketing campaign for the greenhouse business. His enthusiasm for improving the family business was not shared by his parents, however—to Tupper's disappointment.

THE BUDDING INVENTOR

Tupper began experimenting with various inventions at this time. He drew diagrams of his ideas in a notebook that he carried with him everywhere. These ideas ranged from the very modest, such as a type of comb that could be clipped to a belt, to the wildly impractical, such as his plan for a boat that would be powered by a large fish strapped to the bottom of its hull. He tried to promote his ideas to manufacturers, but with no success.

Another of his correspondence courses had been in tree surgery, so around 1930

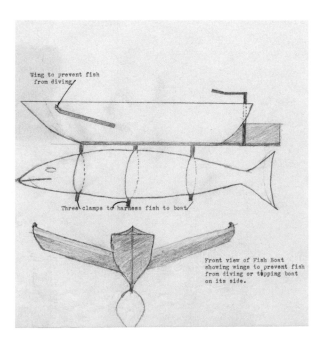

Sketches for one of Tupper's less practical ideas: a boat powered by a large fish.

he began a company called Tupper Tree Doctors. He supported himself in landscaping and tree care for several years. In 1931 he married Marie Whitcomb; the couple would have five children. As the Depression became more severe for greater numbers of Americans, however, Tupper's client list shrank. Tupper Tree Doctors went bankrupt in 1936.

PLASTICS

That failure helped spur Tupper to move in the direction of his true passion: inventing. He had determined as a boy that he would make a million dollars as an inventor by the time he was 30 years old. Although he missed that target by a few years, he was now more certain than ever that inventing was how he would achieve his ultimate success.

In 1937, Tupper met Bernard Doyle, founder of a plastics company in Leominster, Massachusetts. Leominster was becoming known as "plastic city," a manufacturing center where many pioneering inventors and entrepreneurs were establishing new concerns. Doyle's company, Viscoloid, had been bought by DuPont in 1928. Doyle brought Tupper into DuPont as a designer in 1937. At DuPont, Tupper later recalled, his true education began. Although he remained at DuPont for only one year, the time he spent there gave him valuable practical experience in design and manufacturing. It also introduced him to a wide range of contacts who would be instrumental as Tupper set out on his own entrepreneurial ventures.

> I've got that Tupper feeling up in my head Deep in my heart, down in my toes. I've got that Tupper feeling all over me All over me to stay. Yeah!!
>
> —Company song

In 1938 Tupper left DuPont and founded the Earl S. Tupper Company. With some used molding machinery he had purchased, he began making small items such as beads and plastic cigarette containers. Most of the work he did was as a subcontractor to DuPont. During World War II, Tupper's company prospered, largely because of its contracts to produce plastic moldings for gas masks as well as signal lamps for the military. After the war, Tupper refocused his efforts on his first and greatest interest: the civilian consumer market.

A BETTER PLASTIC AND THE TUPPER SEAL

After developing a variety of products such as unbreakable drinking cups (for American Thermos), Tupper fixed his attention on creating a superior plastic container that would keep food fresh for an extended period. American consumers had not taken well to plastic containers in the past, partly because the plastic used had been of an inferior quality and, consequently, often peeled and imparted bad odors to the food. Moreover, no plastic container had succeeded in maintaining freshness;

air easily entered these containers and quickly spoiled the contents. Tupper set out to address both of these shortcomings.

First, in 1938 he developed a superior type of plastic he called "Poly-T." (Tupper insisted throughout his life on the use of the word Poly-T instead of plastic.) Poly-T was a variation of the polyethylene he had worked with at DuPont. Polyethylene—derived from slag, the waste product of refining crude oil—had been used during World War II as insulation for electrical wiring. A black, hard, foul-smelling substance, it was ill-suited for domestic use. Tupper invented a method for refining and cleaning this slag to create a substance that was clear, lightweight, odorless, and hard yet flexible. The resultant Poly-T could bend without shattering, was resistant to heat, imparted no odors to food, and was very inexpensive.

Second, Tupper developed an improved Poly-T sealing lid for the container—his famous burping lid. This seal, based on the lip found on paint can lids, allowed the user to attach the lid to the plastic base, then, while gently pressing the middle of the lid, lift one edge off the lip slightly, thereby "burping," or expelling the air from the container before reattaching the lid over the container's grooved lip. Food kept inside this partial vacuum would stay fresh much longer. Now that he had perfected the composition and the design of his new product, all he needed was a way to show it to the wider public.

A true perfectionist, Tupper oversaw nearly every aspect of his production facility. He demanded the highest standards from everyone who worked for him, he designed every new style of Tupperware, and he looked over the shoulders of the people who ran his machines. Nevertheless, even though he was very driven to succeed, he was unable to bring his products to public attention at first. He initially marketed Tupperware by offering these items as "giveaways" with the purchase of some other product he was manufacturing, such as plastic cigarette cases for Camel. Finally, in 1946, he managed to introduce Tupperware into some hardware and department stores. What became known as the

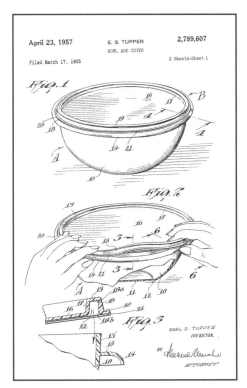

Illustrations from Tupper's patent for his improved lid.

Earl Tupper and
Brownie Wise
visit a
Tupperware
factory in
Farnumsville,
Massachusetts
(undated photo).

"Wonderbowl" gained wider attention and even won some design prizes. However, despite all the marketing efforts, the product was not selling very well, partly because of the bad reputation of previous plastics used for storing food. Both the quality of Tupper's plastic and the "Tupper seal" needed to be demonstrated to customers to create demand.

THE RISE OF TUPPERWARE

Soon after Tupperware began to appear on retail shelves, Stanley Home Products added it to a line of household products that the company

The Tupperware Sales Force

The system of face-to-face marketing and sales that developed around Tupperware products was developed and refined mostly by women. That sales approach in turn opened important opportunities for women at a time when many other avenues remained closed to them.

During the years of American involvement in World War II, women entered the workforce in greater numbers because so many men were serving in the military. When soldiers returned home at the end of the war, they were given many of the jobs that women had held for the preceding years. The women were, in effect, sent back home.

Through its sales approach of holding home parties (Tupperware parties) designed to demonstrate and sell products directly to invitees, Tupperware provided millions of jobs to women. Mothers could schedule their parties and related sales activities around their other responsibilities. Typically, a Tupperware dealer would find other women willing to host events in their houses; these hostesses would invite their friends and family, and in return each hostess would receive a gift of a home product from the dealer. The Tupperware party became as popular a social phenomenon as Tupperware products were popular in the kitchen.

It was not just as dealers of Tupperware that women could enjoy economic success. Those who wanted to branch out could do so by becoming managers of other dealers. Ambitious dealers would always be on the lookout for potential hostesses—and as they branched out, they might also look for other dealers to recruit. As managers, these women would train and motivate new recruits and earn a commission on the events they booked. Successful managers were eligible to become distributors, who were given exclusive rights to distribute Tupperware products in a given region of the country. However, a woman could become a distributor only on the condition that her husband join her in the enterprise, requiring that the husband quit his job. The financial opportunities for distributors during the 1950s and 1960s were such that couples usually accepted this condition. Still, the assumption that women could not fill the top positions in the hierarchy without their husbands' involvement speaks to the bias even within the very system that, in its era, was perhaps most favorable to women.

Brownie Wise, standing right, tosses a bowl during a Tupperware party in the 1950s.

demonstrated at parties held in private homes. Stanley's representatives did well with the product, selling Tupperware in homes much more effectively than it was being sold in retail stores. One agent in particular, Brownie Wise, had been selling Tupperware at a truly impressive rate from her office in Florida, where she purchased the product through local plastics distributors and resold it to the public. Tupper noticed and, in 1948, met with Wise and several other Stanley salespeople to discuss an overall marketing strategy for Tupperware. The salespeople persuaded Tupper that the home-demonstration party was the most effective way to market his product. Tupper adopted a plan based on Stanley's methods, and by 1950 Tupperware was enjoying wider distribution. In 1951, Tupper decided to discontinue supplying retail stores altogether and to rely solely on home-sales parties.

He established a new company, Tupper Home Parties, near Orlando, Florida, and appointed Wise to run it. The production headquarters remained in Massachusetts. In her first full year on the job, Wise managed to double Tupperware sales. The popularity of the home demon-

TIME LINE

1907	1937	1938	1947	1948
Earl Tupper born in Berlin, New Hampshire.	Tupper goes to work for DuPont.	Tupper founds the Earl S. Tupper Company; develops Poly-T plastic.	Tupper begins selling Tupperware.	Tupper meets Brownie Wise, a successful Tupperware home sales agent.

stration party, which would become universally known as the "Tupperware party," grew dramatically. After three years on the job, Wise had increased the number of Tupperware dealers from two hundred to more than nine thousand.

TUPPERWARE'S LEGACY

Earl Tupper became increasingly estranged from the company he founded even as it enjoyed its greatest success. Correspondence between Wise and Tupper indicates that he had grown jealous of Wise for the prominent public role she had assumed as the face of Tupperware. In January 1958, Tupper fired Wise. Later that year, he sold the firm to the Rexall Drug Company for $16 million. Tupperware Home Parties continued to flourish under Rexall, with Tupper serving as chairman of the board of directors until 1973. That year, he moved with his family to Costa Rica, where he became a citizen and lived until his death of a heart attack on October 3, 1983.

Throughout the 1950s, Tupperware became more popular in American homes than any other kitchen product of its kind. Tupperware parties were held in millions of American households to introduce this new product to more and more consumers. By 1960, Tupperware parties were being held in Canada and western Europe. Soon the phenomenon and the product were spreading across the globe. By the end of the 20th century, Tupperware was facing competition from a wide variety of other plastic containers. Nonetheless, Tupperware can be found in nearly every country, with special products developed for particular culinary traditions, such as the "Kimchee Keeper" for sale in South Korea. By 2005, more than 75 percent of the company's $1.3 billion in sales came from outside North America.

The method of direct marketing in homes has continued as the means of selling Tupperware. Although modern women need not rely on the Tupperware party for income as many did in the 1940s, the home party

1951	1958	1973	1983
Tupperware becomes available only through home-sales parties.	Tupper fires Wise.	Tupper retires.	Tupper dies.

One of the Tupperware company's 21st-century lines of products, called Clear Impressions.

concept remains an extremely popular sales tool; companies around the world organize home parties to sell products including toys, jewelry, and gourmet foods. The initial need to demonstrate directly to consumers the qualities of Tupper's Poly-T led to a vast and lucrative marketing approach. This legacy belongs as much to Brownie Wise as it does Tupper. Tupper invented the superior product, but Wise had the vision to recruit and inspire women in the 1950s to be both the sales force and the market that allowed Tupperware to permeate American kitchens and culture.

—Paul Schellinger

Further Reading

Books

Clarke, Alison J. *Tupperware: The Promise of Plastic in 1950s America.* Washington, DC: Smithsonian Institution Press, 1999.

Sparke, Penny, ed. *The Plastics Age: From Bakelite to Beanbags and Beyond.* Woodstock, NY: Overlook, 1993.

Web site

PBS, American Experience: Tupperware, Building an Empire Bowl by Bowl Biography of Earl Tupper and social history of Tupperware. http://www.pbs.org/wgbh/amex/tupperware/

See also: Appert, Nicolas-François; Household Inventions.

ALESSANDRO VOLTA

Inventor of the battery

1745–1827

From pocket calculators and personal stereos to cell phones and laptop computers, almost every compact modern device is powered by batteries. These small, portable reserves of electricity were first invented in 1800 by Italian physicist Alessandro Volta. For much of Volta's life, people believed electricity was something magical produced inside the bodies of animals. Volta's careful research helped to dispel this myth and turned electricity from a scientific novelty into a practical everyday technology.

EARLY YEARS

Alessandro Volta was born in Como, Italy, on February 18, 1745, to an aristocratic family who had fallen on hard times. He showed no signs that he would grow up to be a scientific genius; he did not talk until the age of four, prompting his family to believe that he might be slow. When Volta was just seven, his father died. Young Volta lost himself in books and learning, and he was soon far ahead of other children.

Volta's uncle took responsibility for his education. Between the ages of 13 and 15, Volta attended a Jesuit school, studying such traditional subjects as philosophy and Latin. His uncle intended him to become a priest and enrolled him in a seminary, but Volta had already decided he wanted to be a scientist. Electricity was something he found particularly fascinating; his first contribution to this new field was a Latin poem on the subject.

By the age of 18, Volta was writing excited letters to great scientists, explaining his own theories about electricity and asking for their thoughts in return. He was soon paying frequent visits to the home of Italian scientist Giulio Cesare Gattoni (1741–1809), who let Volta use his laboratory. Gattoni taught Volta physics and helped to propel him toward a scientific career.

In 1769, Volta published his first scientific paper, "On the Forces of Attraction of Electric Fire." In it, he put forward one of the first comprehensive theories of electricity. That paper established Volta's reputation and gained him his first major appointment— as a physics teacher at a high school in Como—in 1774.

ELECTRICITY BEFORE VOLTA

Undated painting of Alessandro Volta with two of his inventions: the battery on the left and the electrophorus on the right.

Electricity is used in commonplace technology in the 21st century, but it was viewed in a completely different way in Volta's time. By the 18th century, people had already known about static electricity (electricity that builds up in one place) for around two thousand years. Around 600 BCE, the ancient Greek philoso-

pher Thales discovered that a rod rubbed with fur had the power to attract other light-weight objects, such as pieces of paper. Later, scientists realized that this was the result of the rod's gaining an electric charge, or a buildup of static electricity. The rod attracted the paper because the paper had an opposite electric charge.

By the early 18th century, electricity had attracted the attention of many great minds, but it was still something of a mystery. In 1733, Charles-François de Cisternay Du Fay (1698–1739) published a paper in which he described two types of electricity—he termed them *vitreous* and *resinous*—that are now known as positive and negative charges. In the 1750s, the American inventor and statesman Benjamin Franklin (1706–1790) began a more scientific study of electricity. Franklin refined a theory that others had put forward: electricity was a kind of "fluid" that could flow from place to place.

In those days, static electricity was the only kind of electricity people understood. Scientists who wanted to study electricity first had to make some static electricity to study. Not all of them were like Franklin, who famously (and very dangerously) flew a kite in a thunderstorm to gather static electricity from clouds. Most, including Franklin, used a Leyden jar for their experiments. A Leyden jar—a special kind of glass and metal container that could store static electricity—was the first capacitor and a very early forerunner of the battery.

INVESTIGATING ELECTRICITY

In 1775, after studying Franklin's work, Volta invented a simple device, an electrophorus, that enabled scientists to make small electrical charges for their experiments. It worked a bit like a rubber stamp. The electrophorus had two parts: a flat base made from a plastic-like substance (equivalent to the ink pad) and a metal disk with a plastic-like handle (equivalent to the stamp). When the base was rubbed to give it an electric charge, the disk with the handle could be brought up to it, then used to transfer the electric charge to other objects—just as a rubber stamp transfers ink. The process could be repeated again and again, so it was easy to perform a series of electrical experiments. Volta's electrophorus was a great success: many scientists began using it instead of the Leyden jar, and it is still used in schools for experiments involving static electricity.

Although the electrophorus was mainly a scientific instrument, Volta could immediately see practical ways of using it. In 1777, he suggested that an electrophorus could be used to make a primitive telegraph system for sending messages using electricity. Although he carried out a number of experiments, Volta never managed to turn this idea into a practical invention. Had he done so, he would have pioneered the electric telegraph around fifty years earlier than its eventual inventor, Samuel Morse (1791–1872).

CHEMICAL RESEARCH

In the 18th century, science was much less specialized than it is today; scientists commonly investigated a broad range of different topics and phenomena instead of confining themselves to a single narrow discipline. Thus during the late 1770s, Volta turned his attention from electricity to the study of chemistry.

Shortly after inventing the electrophorus, Volta decided to vacation at the popular Lake Maggiore in northern Italy. During a boat trip on the lake, he was pulling in alongside the bank and happened to poke a stick into the mud. As he did so, he noticed bubbles of gas coming up from the marshes. Acting like the scientist he was, he collected a sample of the gas and took it back to his laboratory. There, he found that the gas was flammable (burned readily) and named it "marsh gas." It is now widely known as methane, a gas made of carbon and hydrogen, and Volta is credited with its discovery.

ANIMAL ELECTRICITY

In 1778, Volta's achievements led to his appointment as professor of physics at the University of Padua, where he remained for the next 25 years. At Padua Volta became embroiled in a major scientific controversy.

About fifteen years after Volta's appointment Luigi Galvani (1737–1798), an Italian professor of anatomy, made what appeared to be a very surprising discovery about electricity in 1792. When he took the leg from a frog that he had recently killed and touched it with two different pieces of metal, the leg twitched slightly. Galvani explained this strange result by suggesting that the frog's leg contained a kind of

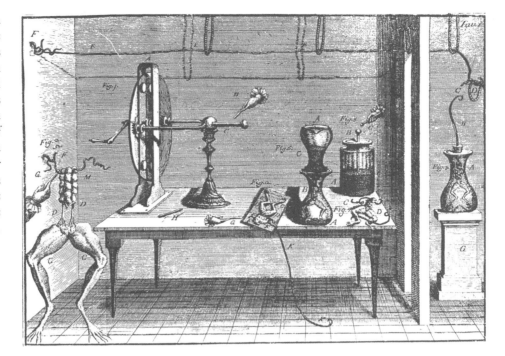

An illustration from Galvani's De viribus electricitatis in motu musculari commentarius *(*Commentary on the Effect of Electricity on Muscular Motion*), which details his experiments on frogs.*

"animal electricity" that made its nerves twitch when he touched it with the metal. Galvani knew that some animals, such as electric eels, naturally produce electricity. However, he used this new discovery to go farther: he suggested that all animals could produce electricity inside organs such as the brain, and this electricity would flow around their bodies through their nerves. Although Galvani's theory of animal electricity was wrong, his idea that electrical impulses flow through the nervous system was a major discovery.

When Volta heard of these experiments, he refused to believe the conclusions about "animal electricity." He thought instead that the electricity was produced by the metals Galvani had used. Volta believed the frog's leg twitched simply because it was part of an electrical circuit (the closed path around which electricity flows). He set about repeating Galvani's experiments to test this idea. Using different animals and different metals, he showed that electricity was produced at the juncture where the two metals met, not in the animal tissue.

Volta called his theory "metallic electricity," but even though his experiments seemed decisive, the controversy between his theory and Galvani's rumbled on for many years. Although Volta and Galvani disagreed in their explanations for what they observed, they were good friends and each respected the other's work. Volta said Galvani's experiment "contained one of the most beautiful and most surprising discoveries"; indeed, Galvani's work was the spur that prompted Volta's next invention.

How Batteries Work

"It is the difference in metals that does it."

In attempting to re-create Galvani's frog experiment, Volta realized that the electricity was not coming from the frog itself. As he wrote, "it is the difference in metals that does it." This realization enabled him to invent the battery in 1800.

Batteries are portable supplies of electricity that make their power through chemistry. A typical battery is a small metal cylinder with two outer terminals, or electrical connections, at the top and at the bottom. Just as in Volta's original battery, the outer terminals are connected to two dissimilar metals inside the case; in Volta's battery, the two metals were zinc and silver. In between the terminals and contained inside the battery case is a chemical called an electrolyte. In most modern batteries, the electrolyte is a dry powder; in others, such as car batteries, it is a liquid.

A battery starts working when it is connected into a continuous electrical loop (circuit). In a flashlight, for example, the circuit consists of the battery, a small lamp, a switch, and some short wires that connect them. When the switch is on, the circuit is complete and electricity flows. This happens because chemical reactions begin to take place inside the battery, causing the electrolyte to slowly

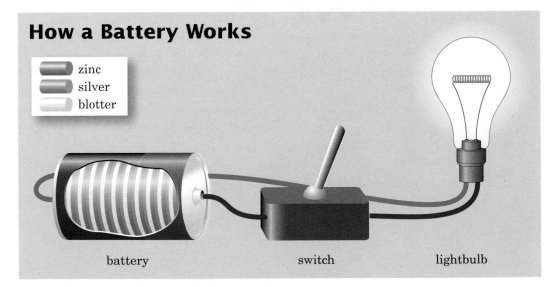

How a Battery Works

zinc
silver
blotter

battery   switch   lightbulb

The battery in this diagram is made from alternating disks made of zinc and silver and cardboard soaked in salt water (called the blotter). Modern batteries may not use these materials but they are based on the same principles Volta discovered.

transform into different chemicals. As it does so, tiny electrically charged particles flow through the electrolyte between the two battery terminals. Electrons (tiny negatively charged particles) flow in one direction; ions (positively charged atoms) flow the opposite way. The moving electrons and ions make the electricity that flows through the battery and out around the circuit, causing the lamp to light.

A battery supplies power until the chemical in the electrolyte is fully transformed; it then runs flat. If it is a rechargeable battery, such as those used in laptop computers, connecting it to an electrical outlet makes the same chemical reaction run backward. As the reaction reverses, the battery takes in and stores energy ready to be used later.

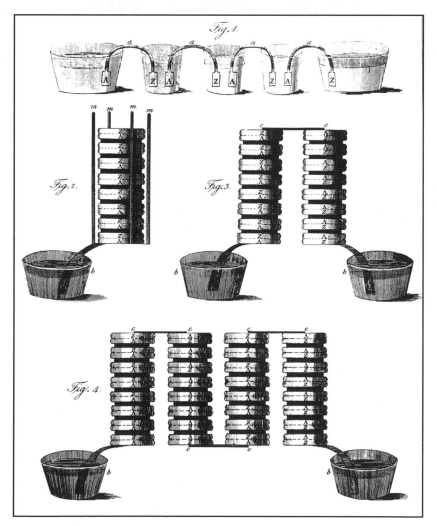

Illustrations of Volta's battery, from an article he published in the British journal Philosophical Transactions of the Royal Society *in 1800.*

INVENTING THE BATTERY

Volta settled the issue of animal electricity in 1800 when he developed the invention for which he is best remembered: the battery. During the course of his experiments on metallic electricity, he was working with different pairs of metals to see which ones produced the most electricity. He found that if he stacked zinc and silver disks and separated them with cardboard soaked in salt water (called the blotter), he could produce a steady, continuous supply of electricity—exactly what scientists needed for their experiments. Originally he called his invention an "artificial electric organ," because it generated electricity in a manner similar to animals such as electric eels. Later, it came to be known as the Voltaic pile because it was literally a pile of metal disks. The modern equivalent of Volta's invention is called a battery.

Volta first announced his invention in a letter to Joseph Banks (1743–1820), president of the London Royal Society, on March 20, 1800; the announcement caused a sensation. In 1801, he was invited to travel to Paris to show the pile to the French emperor Napoleon I. After watching the demonstration, Napoleon was so impressed that he rewarded Volta with a medal and the titles count and senator of the kingdom of Lombardy. Although Volta was grateful for the honors, he apparently took much more pleasure in his family than in his titles. In November 1801, shortly after receiving his award, he wrote to his wife: "Among the many things which indeed give me great pleasure, I do not delight in believing that I am more than what I am: and to a life upset by vainglory I prefer the peace and sweetness of domestic life."

LATER YEARS

From Le Petit Journal Paris, *an illustration of Volta (standing) demonstrating his battery to Napoleon I.*

This letter signaled the beginning of Volta's retreat from scientific work: he lived almost another thirty years, but made no more great discoveries. In 1813,

1745	1769	1774	1775	1778
Alessandro Volta born in Como, Italy.	Volta publishes his first scientific paper.	Volta becomes a high school physics teacher.	Volta invents the electrophorus.	Volta is appointed a physics professor at the University of Padua.

he gave up teaching. That year, he was visited by English chemist Humphry Davy (1778–1829) and Davy's young assistant Michael Faraday (1791–1867), who were on a grand tour of Europe. Davy and Faraday were two of the important scientists who expanded on Volta's chemical and electrical discoveries and inventions during the early decades of the 19th century.

As Volta grew older, he was eager to spend more time with his family. In 1815, however, he was given a new appointment as dean of the faculty of philosophy, to ensure that the University of Padua did not lose his expertise. The following year, his collected works were published in Florence in five volumes. Volta finally retired to a country home in Como in 1819. He died there eight years later on March 5,

Even in the 21st century, researchers continue to find new uses for ordinary batteries; this aircraft, which was presented in 2006 at the Tokyo Institute of Technology, runs on 160 AA batteries.

TIME LINE (continued)

1800	1801	1815	1819	1827
Volta invents the battery.	Volta demonstrates the battery to Napoleon I and receives a medal.	Volta becomes dean of the faculty of philosophy at the University of Padua.	Volta retires.	Volta dies.

1827, at age 82. Fifty-four years after his death, in 1881, Volta's achievements were immortalized when scientists named an electrical unit, the volt, in his honor.

At the start of the 18th century, electricity was little more than a curiosity—an interesting phenomenon that had no practical use. By the end of the 19th century, American inventor Thomas Edison (1847–1931) had opened the first coal-fired power plants to supply electricity to the world. In the decades between, scientists such as Benjamin Franklin, Alessandro Volta, and Michael Faraday helped to put the study of electricity on a truly scientific footing, thus making possible Edison's later work and the modern age of electric power.

—Chris Woodford

Further Reading

Books

Parker, Steve. *Eyewitness: Electricity*. New York: Dorling Kindersley, 2005.
Woodford, Chris. *Routes of Science: Electricity*. Farmington Hills, MI: Blackbirch, 2004.

Web sites

Electricity Timeline
   An interactive history of electricity.
   http://www.schoolscience.co.uk/flash/electric/index.htm
Welcome to Alessandro Volta
   An Italian museum dedicated to Volta's life and works.
   http://ppp.unipv.it/Volta/

See also: Edison, Thomas; Energy and Power; Faraday, Michael; Franklin, Benjamin; Morse, Samuel; Science, Technology, and Mathematics.

ROBERT WATSON-WATT

Pioneer of radar

1892–1973

Science and technology often develop most quickly in wartime. During World War II, thousands of scientists turned their attention and knowledge to developing military inventions that could help their nations in battle. One of the most effective inventions—radar—was pioneered by British scientist Robert Watson-Watt. Radar is best known as a way of using invisible radio waves to help ships and airplanes navigate. However, it also has many other uses, from space research to weather forecasting.

EARLY YEARS

Robert Watson-Watt was born on April 13, 1892, in the town of Brechin, Aberdeenshire, in Scotland. His father was a carpenter, as his grandfather had been. His mother was a passionate feminist and was also active in temperance reform, a worldwide movement that tried to reduce widespread alcohol abuse in the 19th century.

After attending a local school in Brechin, Watson-Watt went to the University of Dundee to study electrical engineering. It was a new and exciting field, particularly since the Italian inventor Guglielmo Marconi (1874–1937) had recently made the first practical demonstrations of radio. After Watson-Watt graduated in 1912, he became an assistant to William Peddie, who encouraged him to study radio. In 1915, Watson-Watt moved to the south of England to work for the British government's Meteorological Office. There, he began to develop radio equipment that could be used to detect incoming thunderstorms. He was granted his first patent for a radio detecting device in 1919.

PIONEERS OF RADIO DETECTION

Watson-Watt did not invent the idea of detecting objects by the use of radio waves. Like radio itself, this invention owed much to many different people. For example, British physicist James Clerk Maxwell (1831–1879) set out the basic science of radio in 1873. A German

Robert Watson-Watt conducts an experiment with a kite and a wireless transmitter in 1931.

physicist, Heinrich Hertz (1857–1894), made the first radio waves in his laboratory some years later. The first person to try to detect objects with radio waves was another German, Christian Hulsmeyer (1881–1957). In 1904, he patented the telemobiloscope, a radio device that ships could use to locate one another. The German navy had little interest, however, and the device was never manufactured.

The first real demonstration of the power of radio detection came from scientists who were trying to understand the earth's atmosphere. Marconi had demonstrated that radio waves could travel immense distances around the curved earth, something many had thought impossible. Scientists believed that the earth's atmosphere was responsible: they thought it was bending radio waves around the planet's curved surface, a bit like a mirror in space. In 1924, English scientist Edward Appleton (1892–1965) fired radio waves into the atmosphere to prove this hypothesis; Gregory Breit (1899–1981) and Merle Tuve (1901–1982) conducted a similar experiment in the United States in 1925. These scientists timed how long the radio waves took to return and so proved the existence of the ionosphere, the part of the atmosphere that bends radio waves around the earth. These experiments were the first successful use of radio detection.

FROM DEATH RAYS TO CHAIN HOME

While this pioneering work was under way, Watson-Watt was still working in the Meteorological Office. In 1927, following a government reorganization, his department was merged with another to form the Radio Research Station, and Watson-Watt was put in charge. He was a popular and enthusiastic boss who respected the scientists who were working for him. Teamwork proved vital in the development of radar.

During the 1930s, Germany was enlarging its military forces at a rate that alarmed other European nations, who feared attack. In 1934, Britain conducted a mock air raid on its capital city, London, to see how easily enemy bombers could reach it. Even with air defenses in place, more than half the mock bombers found their targets. Around the same time, rumors reached England that the German leader, Adolf Hitler, was developing a "death ray": an invisible, high-powered beam of radio waves that could be fired from an airplane to obliterate an entire city. Watson-Watt was asked to advise the British government about whether this was possible. His assistant, Arnold F. Wilkins, quickly calculated that radio waves were much too weak for such use. However, Wilkins and Watson-Watt realized that radio beams did have important military potential—as air defense.

On February 26, 1935, they built a primitive test system using a BBC radio transmitter and showed that it could detect a bomber up to eight

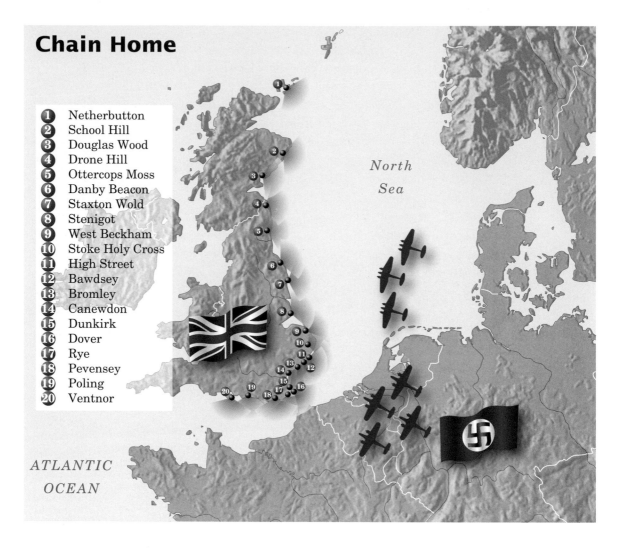

Chain Home

1. Netherbutton
2. School Hill
3. Douglas Wood
4. Drone Hill
5. Ottercops Moss
6. Danby Beacon
7. Staxton Wold
8. Stenigot
9. West Beckham
10. Stoke Holy Cross
11. High Street
12. Bawdsey
13. Bromley
14. Canewdon
15. Dunkirk
16. Dover
17. Rye
18. Pevensey
19. Poling
20. Ventnor

North Sea

ATLANTIC OCEAN

The map above shows the 20 stations of the Chain Home radar defense system, used by Great Britain during World War II.

miles (13 km) away. The government awarded them £10,000 for further research (equivalent to almost $2 million today) and they built a research station on the east coast of England at Orford Ness. By June of that year, their Radio Direction Finding (RDF) system, as it was then known, could detect ships and planes approaching more than 40 miles (64 km) away—around 11 minutes before they posed a serious threat. Between 1935 and 1939, Watson-Watt and his team built a series of 20 radio detection stations around the south and east coasts of England, named Chain Home. This original network was later doubled in size.

THE WORLD AT WAR

By September 1938, Chain Home was operating 24 hours a day. When war broke out in 1939, the system proved invaluable in helping Britain defend itself against air attacks from Germany. Watson-Watt's radar

How Radar Works

People can see things because light reflects off objects and makes them visible. Light travels 186,000 miles (299,338 km) in a second, so perception happens more or less instantaneously. To see at night, cars shine their headlights into the darkness. Light travels out from the car, hits objects in the distance, and reflects back into the driver's eyes. Headlights are ineffective in airplanes and ships, however, because darkness and bad weather can sometimes prevent pilots from seeing more than a few feet ahead.

Radar is like an invisible headlight that can cut through darkness and bad weather. Inside an airplane the radar transmitter regularly sends out coded signals consisting of radio waves. These are similar to light and they travel just as quickly, but they pass through darkness and fog and travel much farther. The radio waves continue to travel until they hit something (for example, a nearby helicopter), then bounce back again. The airplane also has a radar receiver that detects incoming radar signals.

When it receives the signals, it decodes them and calculates how long they took to travel to the helicopter and return. Knowing that the signals traveled at 186,000 miles (299,338 km) per second, it can figure out how far away the helicopter is and how quickly it is approaching. Pilots can use this information, like a second pair of eyes, to alter course and avoid a collision.

The Radar Approach Control Center of the 1961st Communications Group, part of the U.S. Air Force, photographed in 2005.

research team continued to refine and improve radar apparatus. In 1940, two members of the team, Henry Boot (1917–1983) and John Randall (1905–1984), developed the magnetron, a piece of apparatus that generated radio waves at higher power and frequency. Radio detecting equipment could now be made smaller and carried more easily on airplanes and ships. Thus it could be used not just to defend against enemy bombers from the ground but also to help British planes aim their own missiles more effectively. The team was thought to be a target for enemy attacks, however. In 1940, it was moved farther out of range of German bombers to a farm in the remote village of Worth Matravers on the south coast. Two years later, it was moved to safety again—this time inland to the Malverns near Wales.

The British were not alone in developing radio detection for military use. They worked closely with scientists in the United States, who found a much snappier name for the invention. They called it radar (radio detecting and ranging) because it used radio waves to locate objects (detecting) and to find out how far away they were (ranging). The U.S. Army developed powerful radar of its own, both for detecting incoming planes and for controlling antiaircraft guns to stop them. Radar like this was installed in Hawaii in 1941 and detected the approach of Japanese airplanes on December 7 shortly before the attack on Pearl Harbor.

Radar, as it was now known, gave Britain and its allies an advantage over Germany. Ironically, the Germans had developed their own version of radar before the war. Confident of victory over Europe, they canceled its development in 1940. During that same time, the British and Americans stepped up their radar research—and reaped the rewards. In the summer of 1940, the Luftwaffe (the German air force) launched a major air attack on England that became known as the Battle of Britain. Using 2,600 airplanes, Germany tried to devastate the British forces in preparation for an invasion. The British had only

TIME LINE

1892	1915	1927	1935	1942	1973
Robert Watson-Watt born in Brechin, Scotland.	Watson-Watt begins working for the Meteorological Office.	Watson-Watt is put in charge of the Radio Research Station.	Watson-Watt and his team begin building the Chain Home radar system.	Watson-Watt is knighted.	Watson-Watt dies.

a quarter as many airplanes. Even so, they won the air battle using radar, which helped them locate and destroy enemy planes much more effectively.

RADAR IN THE MODERN WORLD

After World War II ended in 1945, radar continued to find important military applications. For much of the late 20th century, the world was engaged in the Cold War: the two superpowers (the United States and the former Soviet Union) built up enormous stocks of long-range missiles to use against each other. Large radar installations were established at the same time to warn each side against impending missile attacks from the other. For example, the United States built—and still maintains—giant early-warning radar systems in remote parts of Alaska, Greenland, and England.

When jet airplanes became popular in the 1950s, and the skies grew more crowded, radar became essential for navigation. Air-traffic controllers also started using radar to manage the flow of planes into and out of airports. Similarly, harbormasters use radar to steer ships into and out of busy ports, and ships themselves began to find radar invaluable for navigating in bad weather or when visibility is poor. Radar makes air and sea travel considerably safer than it would be if pilots and captains had to rely entirely on what they could see—as was the case in the early 20th century.

In the modern world, radar has found many uses apart from military detection and navigation. Robert Watson-Watt first used radar

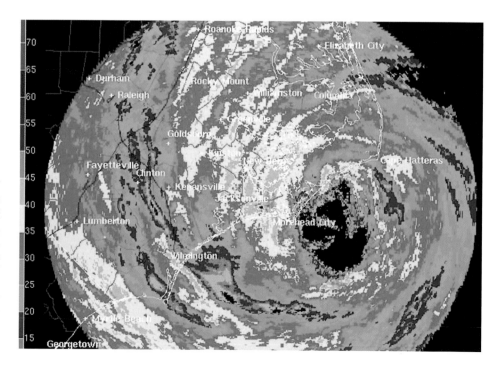

A Doppler radar image from the National Weather Service of Hurricane Isabel centered over Cape Hatteras, North Carolina, on September 18, 2003.

technology to help him detect thunderstorms; a type of radio detector, Doppler radar, is one of the most useful tools for weather forecasters today. It can locate advancing storms and determine both the probability of rain's arriving at a certain time and the expected severity of a storm. Police officers use a similar technology in radar speed cameras, which calculate how fast cars are traveling by bouncing a radar beam off them. Some cars even have tiny radar devices built into the fenders to help drivers park. Space probes use similar technology to help them land slowly and safely on the surfaces of planets like Mars.

These important applications of radar were not invented by Robert Watson-Watt, yet he can take the credit for inspiring many of them. Although he did not invent the idea of radio detection, he proved that it could work on a large scale. Indeed, his work may even have changed history: without radar, Britain and its allies might have lost World War II. In recognition, Watson-Watt was knighted in 1942. Ten years later, the British government awarded him £50,000 as a reward for his work. He spent his later years in Canada and the United States, before returning to Scotland. He died there in Inverness, in the Scottish Highlands, on December 5, 1973.

—Chris Woodford

Further Reading

Books

Adams, Simon. *Eyewitness: World War II*. New York: Dorling Kindersley, 2004.

Bridgman, Roger. *Eyewitness: Electronics*. New York: Dorling Kindersley, 2000.

Web sites

How Stuff Works: How Radar Works

An explanation of the science and technology behind radar.

http://electronics.howstuffworks.com/radar.htm

Radar Pages

A history of British radar and Robert Watson-Watt.

http://www.radarpages.co.uk/

See also: Communications; Marconi, Guglielmo; Military and Weaponry.

CHARLES WHEATSTONE

Prolific inventor

1802–1875

The 19th century was the age of the polymath: a creative person who makes important contributions in many different fields. Charles Wheatstone was a typical inventor of the period. In addition to developing one of the first electric telegraphs, he invented an early microphone, an important mathematical code, a machine for showing three-dimensional pictures, and a series of unusual musical instruments. As a scientist, he made important discoveries about electricity, chemistry, and sound.

EARLY YEARS

The son of a shoemaker and musician, Charles Wheatstone was born on February 6, 1802, near the English city of Gloucester. When he was four, the family moved to London and his father became a flute teacher.

Although Wheatstone attended school in both Gloucester and London, he had little formal education. Nevertheless, he had a huge appetite for knowledge and loved to read. He taught himself several contemporary European languages, as well as Latin and ancient Greek, and he read widely about science. Wheatstone was shy—an aspect of his character that would play an important part in his later life.

MUSICIAN

In 1816, when Wheatstone was 14, he was apprenticed to an uncle who owned a music shop. Music was always one of Wheatstone's joys, and around this time he began to invent musical instruments. The first one, the flute harmonique, was an improved version of the flute his father had played. In 1818, he developed his strangest instrument, the acoucryptophone, which seemed to play itself. It consisted of a lyre (a small, handheld harp) that Wheatstone placed in a room where his audience gathered. The lyre hung from brass wires that passed through the ceiling to instruments in the room above. When these instruments were played, the sound vibrations they made traveled down the brass wire and made the lyre play also—as if by magic.

When Wheatstone's uncle died in 1823, Charles and his older brother William took over the music shop. Charles gained a new opportunity to investigate the science of making music. He carried out many experiments and investigated how different sounds traveled when they made air and other materials vibrate. He believed that, under the right conditions, sound would travel 200 miles (322 km) or more. He looked forward to a time when music might be piped into people's homes like gas.

During this time Wheatstone invented a kind of microphone. Modern microphones use magnetism to convert sound waves

Charles Wheatstone, photographed around 1870.

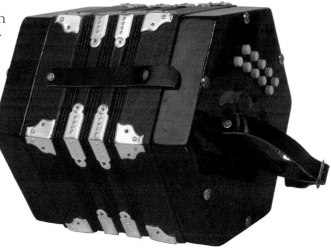

The concertina was invented by Wheatstone in 1829 and has been used by many folk musicians since then.

into electricity, which can then be recorded or amplified. Wheatstone's microphone was very different: it consisted of two solid rods pushed against a person's ears. The rods picked up sounds from the air and amplified them by carrying the sounds directly into the head.

In 1829, Wheatstone began developing what would become his most famous instrument, the concertina. A much smaller version of the accordion, the concertina used a pair of hand-operated bellows. It was an immediate success, especially with the upper and middle classes, who considered it an amusing new drawing-room entertainment. In the 21st century, the concertina is a popular instrument and has been taken up by some folk musicians. The brothers' company, Wheatstone & Company, was soon known exclusively as a maker of concertinas. However, Charles Wheatstone gradually withdrew from it, leaving his brother to manage the firm while his own life headed in a new direction.

SCIENTIST

Wheatstone's research into sound had brought him to the attention of other scientists, including Michael Faraday (1791–1867), a chemist and physicist then working at London's Royal Institution. Wheatstone's experiments helped to inspire Faraday's most famous invention: the electricity generator. When Wheatstone showed how sound could be "induced" to travel between two places, Faraday realized that electricity might travel this way, too, by a process he called "electromagnetic induction." Faraday and Wheatstone became close friends; Faraday even delivered lectures on Wheatstone's behalf because Wheatstone was too shy to talk to crowds of people.

Wheatstone's brilliance was recognized in 1834, when he was offered a job as professor of experimental physics at King's College, London University. In an instant, the amateur experimenter became a professional scientist. One of his first experiments in his new job involved measuring the speed at which electricity travels. The apparatus he invented was later used by other scientists to make the first reliable measurements of the speed of light. The following year,

Wheatstone showed that sparks made from different metals generate light of different colors, so the color of a spark can be used to identify an unknown metal.

INVENTOR

In the 1830s, Wheatstone joined forces with another inventor, William Cooke (1806–1879). Together, they developed one of the first electric telegraphs, which sent written messages using pulses of electricity. The Cooke–Wheatstone telegraph, invented in 1837, originally used six wires to transmit signals, which were sent from and received by large wooden boxes with metal needles that spun around to point to different letters of the alphabet. It was a great success in Britain and was widely used there until the 1870s. However, the simpler telegraph that had been patented in 1837 by Samuel Morse (1791–1872) was less expensive to construct and went on to become much more successful in the United States and other countries.

The telegraph made Wheatstone a household name during his lifetime. However, in modern times he is more often linked with another piece of electrical equipment—the Wheatstone bridge. This device was actually invented by English scientist Samuel Hunter Christie (1784–1865), but Wheatstone popularized it and the invention now bears his name. It is a simple, diamond-shaped electrical circuit that can be used to measure the resistance of an electrical component (how much electric current it allows to pass through).

The Cooke–Wheatstone telegraph, invented in 1837. Very complex compared with Samuel Morse's system, the Cooke–Wheatstone telegraph had five needles that spun to indicate letters (only 20 letters could be used), while six wires carried messages between stations.

During his years at King's College, Wheatstone invented many additional devices. In 1838, he developed the stereoscope, an instrument that could be used to view two side-by-side pictures so that they fused to form a single, three-dimensional image. In 1840, he developed an electricity generator that was more reliable than Faraday's. The same year, he made a chronoscope—an instrument for measuring extremely short periods of time—as well as an electromagnetic clock. During the

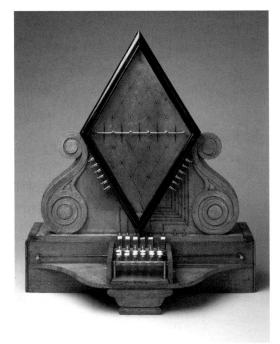

1850s, he developed a ticker-tape system that could be used to send and receive messages by telegraph more quickly and a typewriter that could print onto long strips of paper. One of his most unusual inventions was the Playfair cipher, a mathematical code. This was a five-square grid of letters that could be used to change any sentence into a coded message; it was widely used by the British military until the early decades of the 20th century.

LATER YEARS

Wheatstone received many honors and awards, including a knighthood in 1868. He continued to develop new inventions into his seventies. While on a trip to Paris, France, to promote his latest telegraph inventions, he developed bronchitis and died there on October 19, 1875.

Catching a Killer

Like Samuel Morse in the United States, Charles Wheatstone and his partner William Cooke laid their telegraph cables alongside railroad tracks. In 1845, one of these railroad telegraphs made Charles Wheatstone famous when it was used to catch a murderer.

Early one morning, a woman's body was found in a house north of London. When the police investigated, they discovered that their chief suspect had booked himself on a train to London's Paddington Station in a bid to escape. Knowing that the man was dressed as a Quaker, the local police tried to send a description to their London colleagues using Wheatstone's telegraph line. Unfortunately, Wheatstone's system could not transmit the relatively uncommon letter Q, so the local telegraph operator sent the word "kwaker" instead of "Quaker," causing great confusion. Eventually, he managed to transmit the message: "He is in the garb of a kwaker, with a brown coat on, which reaches nearly to his feet."

When the London police finally understood the message, they rushed straight to Paddington Station and caught the suspect soon after he arrived. The public's imagination was captured by an invention that could send words faster than an express train—and Wheatstone's fame was ensured.

TIME LINE

1802	1816	1823	1829	1837	1868	1875
Charles Wheatstone born near Gloucester, England.	Wheatstone begins inventing musical instruments.	Wheatstone begins experiments with sound.	Wheatstone invents the concertina.	Wheatstone and William Cooke invent an electric telegraph.	Wheatstone is knighted.	Wheatstone dies.

From the time when he bought books as a boy to the day he died, curiosity and ingenuity drove Charles Wheatstone and inspired his achievements. He never really saw himself as a musician, a scientist, or an inventor: he was a polymath, ranging freely over many fields. His huge creative output as both a scientist and an inventor came from putting together ideas from different areas to make important breakthroughs.

—Chris Woodford

Further Reading

Books
Bender, Lionel. *Eyewitness: Invention*: New York: Dorling Kindersley, 2000.
Hart Davis, Adam. *What the Victorians Did for Us*. New York: Headline, 2001.

Web site
BBC History: Charles Wheatstone
> A biography of Wheatstone and essays on other Victorian scientists.
> http://www.bbc.co.uk/history/historic_figures/wheatstone_charles.shtml

See also: Communications; Energy and Power; Faraday, Michael; Morse, Samuel.

ELI WHITNEY

Inventor of the cotton gin

1765–1825

Inventors can change lives—they can even change history. When Eli Whitney invented a machine, the cotton gin, for processing cotton plants, he changed the future for millions in the American South. Later, when he invented a better way of making rifles, he pioneered a manufacturing system called mass production that most factories have used ever since. His ideas were simple, but they could hardly have had a greater effect on the 19th-century history of the United States.

EARLY YEARS

Eli Whitney was born on December 8, 1765, in Westborough, Massachusetts. Although he was the son of a farmer, he preferred tinkering in the workshop to working in the fields, and his great mechanical skill soon became obvious. At the age of eight, he repaired his father's watch; by the age of 12, he was building and repairing violins. Within a few years, he invented a machine for making nails and set himself up in business.

The good-natured teenager was six feet (1.8 m) tall, was broad-shouldered, and had large hands—perfect for farmwork. Whitney, however, longed to go to college, but his family was poor and he had to earn the money for his studies. Between the ages of 18 and 24, he worked as an elementary school teacher, earning seven dollars a month. In 1779, using the money he had saved and with help from his father, he entered Yale College (now Yale University) to study law. To earn money, he continued his nail-making business, made hat pins, and even repaired a valuable orrery (a complicated clockwork model of the planets) belonging to the college president.

MOVING TO GEORGIA

Graduating from Yale at age 26, he was penniless and unsure what the future held. He was offered a teaching job in South Carolina, but the employer changed his mind and refused to pay Whitney as much as they had agreed. When another job came up in Georgia, Whitney made the long journey south—only to find that this offer had fallen through as well. Then he met Catherine Littlefield Greene (1753–1814), the wealthy widow of General Nathanael Greene, a hero of the American Revolution. She took pity on Whitney and invited him to stay at her home, Mulberry Grove, a cotton plantation near Savannah, Georgia.

One evening in 1793, Catherine Greene and her guests were discussing the cotton business. England's demand for cotton was great because of the invention of spinning and weaving machines, but cotton growers in southern states like Georgia could not satisfy the need.

Engraving of Eli Whitney from around 1810.

A cotton plant ready for harvest.

The problem was that green-seed (or short-staple) cotton was difficult to process because the cotton fibers, the part of the plant used to make cloth, stuck to the seeds. Separating the seeds from the fibers by hand took a long time, making the product uneconomical. A different kind of cotton called black-seed (or long-staple) was easier to process because the seeds and fibers did not stick together. Black-seed cotton, however, was hard to grow inland; it grew well only near the coast.

As Catherine Greene and her guests discussed this problem, she challenged Whitney, who had impressed her with his skills by fixing things around her home, to build a machine that would clean green-seed cotton. Perhaps it was just a joke among friends around the dinner table, but Whitney took up the challenge.

THE RISE AND FALL OF THE COTTON GIN

Although Whitney had never seen cotton plants, he soon figured out how to solve the problem: he needed to make a comb that could pull the cotton fibers away from their seeds. Within 10 days, he had built a working model of a cotton-combing machine. Catherine Greene and her young plantation manager, Phineas Miller, were impressed—Miller so much so that he offered to go into business with Whitney to manufacture the machine.

Whitney had studied law, so he knew about the new Patent Act of 1793. This allowed inventors to use a patent to protect "any useful art, manufacture, engine, machine, or device, or any instrument thereon not before known or used." A patent was a drawing with a written description of an invention that went on public record. Taking out a patent gave inventors 14 years to make money from their ideas before the ideas became freely available for others to copy.

By 1794, Whitney and Miller acquired their patent and put the cotton gin into production at a factory in New Haven, Connecticut.

The Cotton Gin

". . .I shall make a fortune by it."

A few pieces of wood and metal fastened together—it is hard to imagine that such a thing could make fortunes, help spark a civil war, and change the history of a nation. These simple components of Eli Whitney's cotton gin, a machine that could split cotton plants into their useful parts, did all this.

Cotton plants grow to roughly a foot (.3 m) high in about two months, when white blooms form. When the flowers fall off, they leave behind a small green pod called a boll. The boll continues to grow and after another six to eight weeks is roughly the size and shape of an egg. Inside are several compartments, each containing perhaps 10 seeds and each seed sprouting thousands of fluffy cotton fibers. Most of the world's cotton is the green-seed (short-staple) variety, in which the seeds and the fibers stick closely together. To make useful cotton for textiles, the seeds and the fibers must be separated.

Before Eli Whitney invented the cotton gin, this job was done by hand, with an entire day needed to separate the seeds from just one pound (.5 kg) of cotton fiber. Using Whitney's gin, a cotton worker could clean the seeds from the fibers by stuffing raw cotton into one end of the machine and turning a handle. Whitney's machine made a dramatic improvement: in a single day, it could clean 50 times as much cotton as 50 people working by hand. Whitney wrote to his father: "Tis generally said by those who know anything about it that I shall make a fortune by it."

Others copied Whitney's invention and cotton gins started to vary in design, but they all worked in broadly the same way with a comb and hooks on a rotating cylinder, just as Whitney had designed them. Small gins were hand-cranked; larger ones, driven by mules or water wheels, were soon in operation; steam-powered gins followed soon after. Although today's gins are huge, factory-sized machines powered by electricity, they still work on a principle similar to the one Whitney discovered.

Workers load cotton into a modern cotton gin in Tchula, Mississippi, in 2004. Modern gins work in the same general way as Whitney's original invention.

How Whitney's Cotton Gin Worked

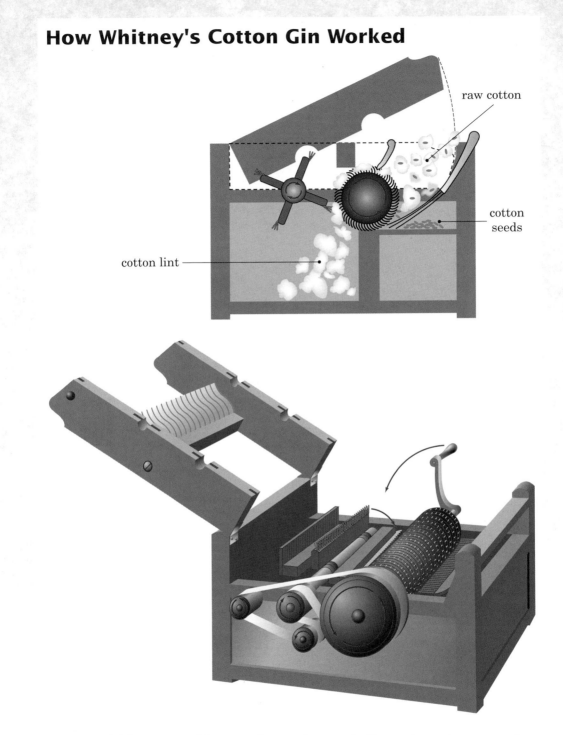

raw cotton

cotton seeds

cotton lint

Cotton was fed down a curved hopper and onto a large comb. Next to the comb was a wooden cylinder with hundreds of metal hooks. As the handle turned, the hooks snagged on the cotton fibers, pulling them upward toward a second comb with narrow teeth. The cotton seeds, which were too big to pass through, dropped back through the wider teeth of the first comb. A second cylinder with bristles on it brushed the cotton fibers off the hooks and into a collection bin. The two useful parts of the cotton plant—the fibers and the seeds—thus ended up in different parts of the machine.

Obstacles arose almost immediately. If Whitney had manufactured his machine the traditional way, each worker would have made an entire gin, but he wanted each worker to assemble only one small part of the gin before passing the machine on to another worker to make the next part. This early version of the modern assembly line was quicker for Whitney, but much less interesting for the workers. Many were craftsmen who resented this way of working, and some quit. Epidemics of scarlet fever and yellow fever then struck New Haven, claiming 114 lives and reducing Whitney's workforce. On March 11, 1795, fire broke out in Whitney's factory and destroyed almost the entire building.

By now, plantation owners had started growing green-seed cotton, and the demand for Whitney's new machine increased daily. With their factory plagued with problems, Whitney and Miller could not make all the cotton gins they had promised. They had agreed not to sell their machine but to rent it to the growers for a 30–40 percent cut of their cotton profits. They did not realize just how much green-seed cotton would be grown and that they could have settled for a much smaller cut and still become immensely rich. Also, the growers thought they were greedy and refused to pay so much. They realized how simple Whitney's machine was and started making their own. Whitney tried to stop them from violating his patent, but a loophole in the patent law left him powerless. Too late, Whitney and Miller drastically cut the price of their gin. By 1797, the two enterprising partners were deep in debt and they went out of business.

Undated illustration of a large-scale cotton gin of the type used on many southern plantations.

In 1798, green-seed cotton was growing throughout the South and the growers were getting rich. Whitney had made all this possible, yet he was penniless and his patent seemed useless. When the patent law changed in 1800, the partners were able to take new legal action. As a result, after 1802, the states of South Carolina, North Carolina, Tennessee and, finally, Georgia were forced to pay to use the cotton gin. Whitney and Miller fought 60 separate suits by 1807 and earned around $90,000—a huge sum in those days—but the money was eaten up by debts and legal costs. When Whitney's 14-year patent on the cotton gin expired in 1807, there was more bad news: Congress refused to renew it. Whitney commented bitterly, "An invention can be so valuable as to be worthless to the inventor." He would never patent anything again.

MAKING MUSKETS

By now, Whitney had a new business. In the late 1790s, he had decided to seek a different way of making money. First, he proposed printing stamps for the U.S. government. When that idea fell through, he thought of making muskets. Many muskets were imported from Europe, but as America neared war with France in the 1790s, the U.S. government realized that it needed to manufacture its own weapons. Although an enormous armory (arms-making plant) had been set up in Springfield, Massachusetts, in 1794, making guns was a slow process: each weapon was made laboriously by one worker. Whitney thought he could speed production using the methods from his cotton-gin factory. In a bold and risky move, he persuaded the government to sign a contract with him to supply 10,000 muskets for $134,000 in just 28 months. At this point, he had no factory, had no workers, and had yet to produce a musket.

Two years later, Whitney had still not produced so much as a single musket; it took him more than 10 years to complete the contract. During that time, he revolutionized the way guns were made. At the armory he built in Hamden, Connecticut, two miles (3.2 km) outside New Haven, he introduced a production-line system in which many unskilled workers each made part of a musket before passing it on to the next person. Instead of being individually handcrafted, all the guns were made to a standard design so the parts from one gun could fit any other. Whitney was one of the first to manufacture goods with interchangeable parts. Instead of using human or animal power, he sited his factory near the Mill River and used water power to operate his machines. Learning from his experience with unhappy cotton-gin workers, he turned his factory into a friendly community called Whitneyville. Apart from providing housing and food for the musket makers, he offered education and training to their children.

LATER YEARS

Whitney became rich making muskets, not as the inventor of the cotton gin as he had hoped. Even so, much of his life seemed to have passed him by: he was already in his twenties when he became a student; legal battles over the cotton gin exhausted him during his thirties; and his forties were spent getting his arms factory off the ground.

In letters he wrote to his old friend Catherine Greene, he spoke of being a "solitary old bachelor." He married Henrietta Edwards, the granddaughter of famous preacher Jonathan Edwards, in 1817. His two nephews, Philos and Eli Whitney Blake, took over running his factory in 1820, and the same year Whitney and Henrietta had a son. By now, Whitney was in his mid-fifties and in poor health. He died five years later, on January 8, 1825.

WHITNEY'S IMPACT

During the 19th century, Whitney's cotton gin changed the face of agriculture in the United States because it made green-seed cotton economical and increased enormously the amount of cotton planted and harvested. By 1800, in the five or six years after the cotton gin was invented, cotton production had increased 25-fold and in every decade thereafter production doubled. In 1850, the United States was growing around three-quarters of all the world's cotton. This cotton brought huge prosperity to the southern states east of the Mississippi River, which grew around 60 percent of the nation's entire crop. Seeing the staggering profits to be made, farmers switched from other crops such as rice and tobacco. The "cotton belt" (the cotton-growing area of the United States) also expanded greatly. Today, it stretches across the South from coast to coast (although little cotton is grown in the states of Florida, Nevada, and Virginia).

Whitney's cotton gin also had a dramatic effect on the history of the United States by breathing new life into the slave trade. When Whitney invented his machine in 1793, slavery was almost a thing of the past. Plantation owners were using slaves in only six states, and even though slave labor was cheap, it was still one of their biggest costs. Many owners either had released their slaves or planned to do so; the

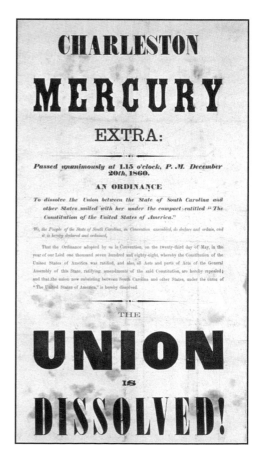

The Charleston Mercury *announces the dissolution of the Union in 1860.*

TIME LINE

1765	1779	1791	1793	1794
Eli Whitney born in Westborough, Massachusetts.	Whitney enters Yale College.	Whitney graduates from Yale.	Whitney develops a working model of the cotton gin.	Whitney begins manufacturing cotton gins.

slave trade seemed to be gradually dying out. Then Whitney's cotton gin appeared, making growing cotton immensely profitable throughout the South. The gin did away with the need for some human labor because it automated cotton cleaning, but more people were needed to pick cotton from the fields. In the 20 years following the invention of the gin, 80,000 slaves were imported from Africa to work on the southern plantations. Congress banned the importing of slaves in 1808, but slavery continued to flourish. By 1860, the number of states using slave labor had grown from 6 to 15 and around one-third of all the people in the South (nearly four million people) were slaves.

Although the Civil War had many causes, slavery was one of the most important: people in the North believed slavery should be abolished because they saw it as cruel, immoral, and inhuman; those in the South wanted to preserve slavery because they saw it as essential to the South's prosperity and as part of the region's culture. A crisis was reached in 1860 after the election of Abraham Lincoln to the presidency on an antislavery ticket. The Civil War between the North and the South began the following year. Around 260,000 southerners were killed, many by mass-produced muskets made at the Whitney armory in New Haven. Whitney's cotton gin had brought prosperity to the South; now his rifles brought the region defeat.

Although he revolutionized agriculture and his inventions helped to start and finish a war, Whitney's greatest contribution to history was still probably his manufacturing process. He thought of a new way of making things using unskilled workers to mass-produce complicated products, such as muskets, from many interchangeable parts. His industrial methods had an even bigger impact when they were taken up by such figures as automaker Henry Ford (1863–1947), pioneer of the mass-produced, affordable automobile. They are still used by most of the world's factories.

—Chris Woodford

TIME LINE (continued)

1795	1797	Early 1800s	1807	1825
A fire destroys most of Whitney's cotton gin factory.	Whitney goes into debt and out of business.	Whitney pioneers the technique of mass production, manufacturing muskets for the U.S. government.	Whitney's patent expires.	Whitney dies.

Further Reading

Books

Green, Constance. *Eli Whitney and the Birth of American Technology*. New York: Pearson Education, 1997.

Mitchell, Barbara. *Maker of Machines (A Story about Eli Whitney)*. Minneapolis, MN: Lerner, 2004.

Patchett, Kaye. *Eli Whitney: Cotton Gin Genius*. Farmington Hills, MI: Blackbirch, 2003.

Web sites

Eli Whitney Museum and Workshop
    Eli Whitney museum in Hamden, Connecticut.
    http://www.eliwhitney.org/
Eli Whitney's Cotton Gin
    An animation showing how the cotton gin worked.
    http://www.eliwhitney.org/cotton/patent.htm

See also: Cloth and Apparel; Ford, Henry; Military and Weaponry.

FRANK WHITTLE

Inventor of the jet engine

1907–1996

Frank Whittle's jet engine vastly increased the speed that airplanes were able to achieve. Faster airplane travel has made the transport of people and goods to faraway places possible, boosting the worldwide economy and expanding people's perceptions of the world. Whittle had enormous trouble getting support to build the jet engine, however, and once it was built, he lost control of his invention.

EARLY YEARS

Frank Whittle was born in 1907 in Coventry, England, where his father worked as a machinist. Whittle occasionally worked in his father's shop, but from an early age he was fascinated mainly by airplanes.

When Whittle was 15 years old, he decided to join the Royal Air Force (RAF) and become a pilot. Since his family could not afford to send him to a training academy, Whittle applied to join the RAF as an apprentice. RAF apprentices were trained to fix airplanes, not to fly them, but a select few were allowed to move on to pilot training. Whittle failed the medical exam, however, because he was too small.

After the exam, a sympathetic RAF sergeant gave him a diet and exercise regimen, which Whittle followed carefully. He grew three inches (8 cm) over the next six months and asked to have his rejection overturned, only to be denied. Whittle decided that his earlier rejection was counting against him, so he applied to the program again, this time pretending that he had not applied before. The ruse worked, and in September 1923, Whittle was accepted into the RAF apprentice program.

Whittle graduated from the program in 1926, sixth out of a class of 600. That year, the top five from his apprentice program were offered the opportunity to go to officer school to become RAF pilots. Whittle seemed to have just missed his chance to be a pilot, but one of the top five had poor vision and failed his medical exam, and Whittle was allowed to go to officer school in his place.

AN IDEA FOR AN ENGINE

To graduate from officer training, Whittle had to write a thesis. He chose as his topic the future of aviation, examining the limitation of the propellers used at the time to power airplanes.

Propellers rotated and pulled air back to push the plane forward. For this reason, propellers did not

A portrait of Frank Whittle from 1948.

How Jet Propulsion Works

By 1929, Whittle was convinced that jet propulsion, which was used to power rockets, could probably work for airplanes. The difficulty lay in designing a practical airplane engine that would use jet propulsion.

Jet propulsion typically works in three stages. First, air is compressed and drawn into a combustion chamber (see diagram). There, the compressed air is mixed with fuel and ignited by a fuel burner. The resulting stream of hot air is ejected out of the back, pushing along whatever needs to be propelled.

The air can be compressed by a device that spins. At first, Whittle thought the compressor would have to be spun by a conventional piston engine, but that design seemed too heavy to be practical. Finally in October 1929, Whittle realized that the hot air streaming out the back of a jet engine could be used to spin the compressor.

He placed a turbine, a windmill-like device, at the back of his engine. A separate starting engine was needed to begin the process, but once the jet engine began to run, the hot air spun the turbine, which spun the compressor, which drew in more air to continue the process. With this design, Whittle had a working jet engine that essentially had only one moving part.

Whittle's Engine

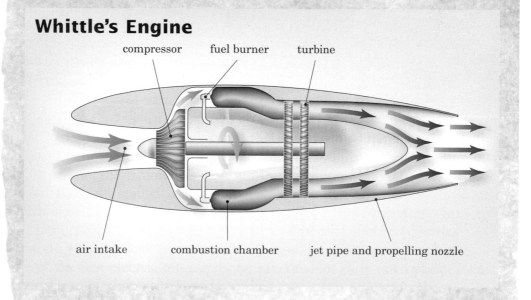

compressor   fuel burner   turbine

air intake   combustion chamber   jet pipe and propelling nozzle

> My experience had given me a clear picture of the special problems peculiar to an aircraft power plant. . . . I saw these things through the eyes of a pilot, as well as through the eyes of an engineer.
>
> —Frank Whittle

work at high altitudes where the air is very thin; the RAF fighter planes of Whittle's day could not fly higher than about twenty thousand feet (6,096 m). That limitation slowed airplanes because thicker air created greater air resistance. In addition, propellers were powered by piston engines, which had hundreds of moving parts. The more powerful piston engines were also bigger, more complicated, and heavier. A heavy engine could potentially prevent an airplane from taking off.

As a result, a typical RAF fighter's top speed was only about 150 miles per hour (241 km/h). In his thesis, Whittle said that some new form of engine would need to be developed that would allow planes to fly higher and faster. Although Whittle suggested a few ideas in his paper, he did not have a complete design. He continued to ponder the issue after he graduated, however, and in October 1929 he developed the basic design of his jet engine.

DISCOURAGEMENT

Whittle submitted his idea to Britain's Air Ministry, which quickly rejected it as impractical. Although Whittle's jet engine was simpler than a piston engine, it would create a tremendous amount of heat, and the ministry doubted that any material could withstand the stress. One of Whittle's friends in the RAF was more optimistic, however, and he encouraged Whittle to patent the engine and to look for private backers.

Concurrently, Whittle had fulfilled his dream of becoming a pilot. He developed a reputation as a daredevil and performed in shows of aerial acrobatics. He also worked for the RAF as a flight instructor and a test pilot. In 1932 the RAF gave Whittle the chance to enter an officers' engineering program. Whittle did so well in the program that the RAF sent him to Cambridge University in 1934 to study engineering.

In January 1935 Whittle's jet-engine patent expired; it required a £5 fee for renewal. Whittle had failed to find a private backer, however, and he became convinced that his jet engine was too far ahead of its time. In addition, by this point Whittle had a family and money was tight. He did not pay the renewal fee, and his patent lapsed.

NEW HOPE

Three months later, Whittle received a letter from two retired RAF officers, one of whom had known him in the service. The two had decided

to go into business together, and they thought that Whittle's engine design looked promising.

The two would still need to find financial backers, but they had better connections to the world of finance than Whittle had. Whittle agreed to join them; in 1936, they found an investment firm, O.T. Falk and Partners, that was willing to fund development of Whittle's jet engine. Whittle established a company to build the engine, Power Jets Limited.

Power Jets arranged with a turbine manufacturer to build its engine, and in April 1937 Whittle's engine was run for the first time. The engine had a tendency to run out of control, but the results were promising enough that the RAF let Whittle work at Power Jets rather than calling him back to military duty. The next year the Air Ministry awarded Power Jets a small contract.

CONTINUING STRUGGLE

The Air Ministry contract proved to be a mixed blessing, however—it was not much money, and the existence of the contract now made the jet engine a state secret. As a result, no one at Power Jets could discuss what the company was doing, and this restriction made finding additional financial backers very difficult.

The strict policy of secrecy was in place because of recent events in Germany, where Adolf Hitler and his Nazi Party were arming the nation for what would become an attempt to conquer Europe. Unbeknownst to the British, German scientist Hans von Ohain in 1933 had independently developed his own design for a jet engine. Unlike Whittle, Ohain had received enthusiastic support. As a result, Germany would fly the first jet airplane in August 1939 and would be the first country to use a jet-powered fighter plane five years later.

Whittle, in contrast, spent the late 1930s scrounging up the money needed to keep his engine-development work going. He was eventually able to build other test engines; in June 1939, his progress persuaded the British government to give Power Jets more support.

WORLD WAR II

In September of that year, Nazi Germany invaded Poland. Great Britain declared war on Germany in retaliation, and World War II was under way. The following year, Germany conquered much of western Europe and launched a massive air attack on Great Britain, which the RAF was barely able to repel.

The British military knew that air strength was vital to the country's defense, so Power Jets received a significant influx of funding. Whittle's engine finally took to the air in 1941, when the *Gloster E28/39* test plane, outfitted with jet engines, made its first flight. The plane reached 370 miles per hour (595 km/h) and an altitude of 25,000 feet (7,620 m).

Despite this success, Whittle was far from happy. Because of the war, the British government had essentially taken over production of his jet engine. The government decided that the Rover Car Company should build the engines that Power Jets developed. Whittle strongly disliked giving trade secrets to a potential rival. His design was not protected by

A Meteor jet fighter, photographed in 1946.

a patent, and Whittle became convinced that once the war was over, Rover's management would enter the jet-engine business with the know-how it had obtained from Power Jets. Since Rover was an established company, it could then crush Power Jets in the marketplace.

The stress, combined with Whittle's growing reliance on drugs to keep him alert during the day and to help him sleep at night, resulted in a serious breakdown. He had to be hospitalized in December 1941—the same month the United States entered the war.

NATIONALIZATION

Whittle left the hospital in early 1942; that summer, the British government sent him to the United States as part of a technology-sharing arrangement. Thanks in part to Whittle's help, the United States was able to produce a jet fighter, the *Airacomet*, which made its first flight in October 1942.

At the same time, the rivalry between Rover and Power Jets was delaying production of a British jet fighter, so in early 1943 Rolls-Royce took over for Rover. Whittle got along much better with the Rolls-Royce management, and the first British jet fighter, the *Meteor 1*, was delivered to the RAF in July 1944. The plane was used to defend Great Britain against German rocket attacks.

Whittle had asked the British government to nationalize the jet-engine industry, putting it under state control to prevent manufacturers from profiting from Power Jets' engine. In January 1944, however, the British government decided to nationalize only Power Jets.

An RAF pilot in the cockpit of a Meteor jet.

TIME LINE

1907	1923	1929	1932	1936	1939
Frank Whittle born in Coventry, England.	Whittle is accepted into the Royal Air Force apprentice program.	Whittle develops a jet engine.	Whittle begins an officers' engineering program.	Whittle founds Power Jets Limited.	Germany flies the first jet airplane.

Although the British government compensated the investors in Power Jets, Whittle considered the amount offered to be insufficient. He refused money for his share in the company because he felt that it was unseemly for a military officer to profit in such circumstances. His unhappiness contributed to another breakdown in March 1944 that required a six-month hospital stay.

AFTER THE WAR

In May 1945, Germany surrendered. Power Jets was now a branch of the British government dedicated to research and development in jet engines. Whittle was still a member of the board of directors of Power Jets, but private manufacturers, to minimize competition, were successful in convincing the British government that a state-owned entity should not develop jet engines.

Power Jets became a research-only operation, and Whittle resigned in 1946. In 1947, he had another breakdown, which resulted in a lengthy hospital stay. The next year, he resigned from the RAF because of poor health.

Whittle would long feel ill-used by the British government, but attempts were made to reward him for his work. Shortly after he resigned from the RAF, the British government awarded him £100,000 and knighted him for inventing the jet engine.

Whittle spent the 1950s consulting for the British Overseas Airways Corporation; in the 1960s, he developed a new kind of drill for oil wells. In 1976, he moved to the United States, where he joined the faculty of the U.S. Naval Academy in Annapolis, Maryland.

Whittle was granted the Order of Merit, one of Great Britain's most prestigious honors, in 1986. In 1991, he shared the Charles Stark Draper Prize of the U.S. National Academy of Engineering with the German jet pioneer Ohain, who had also moved to the United States after the war. Whittle died in Columbia, Maryland, in 1996 at the age of 89.

—Mary Sisson

TIME LINE

1941	1944	1946	1960s	1976	1996
The *Gloster E28/39* test plane makes its first flight.	The *Meteor 1* jet fighter is built.	Whittle resigns from Power Jets.	Whittle develops a drill for oil wells.	Whittle moves to the United States.	Whittle dies.

A worker at the Rolls-Royce aircraft engine factory in Berlin, Germany, with the turbine of a jet engine in 2006.

Further Reading

Books

Golley, John. *Whittle: The True Story.* Washington, DC: Smithsonian Institution Press, 1987.

Jones, Glyn. *The Jet Pioneers: The Birth of Jet-Powered Flight.* London: Methuen, 1989.

Whittle, Frank. *Jet: The Story of a Pioneer.* London: Muller, 1953.

Web sites

Chasing the Sun
    A history by PBS of early aviation.
    http://www.pbs.org/kcet/chasingthesun/index.html

Royal Air Force History: Gloster E28/39 60th Anniversary
    The RAF's exhibit on the first British jet airplane and its inventor.
    http://www.raf.mod.uk/history/e281.html

See also: Goddard, Robert H.; Military and Weaponry; Transportation; Wright, Orville, and Wilbur Wright.

GRANVILLE WOODS

Inventor of devices for
electric railways

1856–1910

Granville Woods grew up in an era when railroads were spreading across the United States and electricity was swiftly changing the nation's cities. The potential of electricity excited and inspired him; he felt he could contribute to the field despite a lack of formal education and widespread discrimination against African Americans. Woods's era was also a time in which business activities were unregulated and scams were extremely common. Woods had far more difficulty managing the business side of inventing than creating his inventions.

EARLY YEARS

Little is known about Granville Woods's early life. Most sources agree that he was born in 1856, but whereas some historians maintain that he was born in Columbus, Ohio, others point to contemporary sources indicating that he was actually born in Australia.

Whatever the case, Woods grew up in Columbus. His family was quite poor, and Woods left school at an early age to work. When Woods was 15 years old, he took a job as a machinist. A year later, he began studying electricity on his own. For the next few years, he continued to work at technical jobs in several cities.

BECOMING AN INVENTOR

Sometime in the mid-1870s, Woods moved to the East Coast (the exact location remains unknown). There, he was finally able to obtain a more formal education, studying mechanical engineering at a technical school for two years. In the late 1870s, Woods returned to Ohio to work for the many railroads there. While at the railroads, Woods became interested in learning to use the telegraph.

Woods also developed an interest in elevators; he decided that the method then in use of communicating between the elevator car and the floors was inadequate (see box, Communicating through Induction). Woods quit the railroads and went to work developing his new elevator-communication system. When he showed his invention to a patent attorney, he was told that it would probably not catch on because of the cost.

Woods decided to develop his induction system for use in railroads. In the summer of 1881, however, he contracted smallpox, a potentially fatal disease that left him unable to work for almost a year.

Woods recovered physically, but the illness left him penniless. He began looking for a financial backer for his induction

Engraved portrait of Granville Woods from the late 1800s.

Communicating through Induction

In the 1880s, Woods invented a system to communicate with moving elevator or train cars. It used a natural phenomenon called electromagnetic induction. Such induction was first observed in 1831, when Michael Faraday (1791–1867) moved a magnet through a coil of wire and discovered that the magnet created an electrical current in the wire.

Woods realized that a magnetic field could be created inside a moving car using an electrical system. The car could then be surrounded by wires, strung up on either side of a railroad track, and the magnetic field would create an electrical current in those wires. Since the electrical system within the car could be turned on and off, the magnetic field could be turned on and off. As a result, the electrical current in the wires outside the moving car could be turned on and off from inside the car. As long as the car was moving, electrical impulses from inside the car could be transferred to wires outside it; the two systems never had to touch.

Radio had not been invented at this point. Without wireless communication, designing a reliable system that would allow people outside a moving car to speak to those inside it was almost impossible. Woods realized that the very motion of the car made communication through induction possible.

Illustration of an experiment conducted on Staten Island, New York, using induction to communicate between train stations and moving trains.

idea and other inventions. In 1883 Woods was able to obtain money from a backer to patent a steam boiler; the next year, he patented an improved telephone transmitter. That year he also patented a device for voice communication over telegraph lines. Woods's patent was eventually purchased by American Bell Telephone Company, which ultimately chose not to develop the technology.

BACK TO INDUCTION

In early 1885 Woods saw an article in *Scientific American* describing an induction communication system invented by a man named Lucius Phelps. The system had already been used on a section of railroad to allow telegraph communication with moving trains.

The article was a shock to Woods, who was concerned that he would lose the chance to profit from his invention. He scrambled to file for a patent on induction telegraphy.

Phelps had filed for the same patent, so the U.S. Patent Office conducted an investigation. Although Woods had filed later than Phelps, he was able to prove that he had originated the idea of using induction to communicate with moving trains in early 1881, well before Phelps.

In 1887, the Patent Office awarded the patent to Woods. Even before Woods won the patent battle, the case had attracted media attention, as well as the attention of some prominent Cincinnati businessmen who became interested in backing the talented inventor. Woods joined forces with them, founding the Woods Electrical Company in 1886.

PROBLEMS WITH PARTNERS

Woods quickly became dissatisfied with his backers. He wanted to develop Woods Electrical into a business, but his backers, who had

TIME LINE

1856	1881	1883	1886	1890	1892	1910
Granville Woods born; grows up in Columbus, Ohio.	Woods works on using induction as a communications tool.	Woods patents a steam boiler.	Woods founds the Woods Electrical Company.	Woods moves to New York City.	A railway system based on Woods's ideas is demonstrated at Coney Island.	Woods dies.

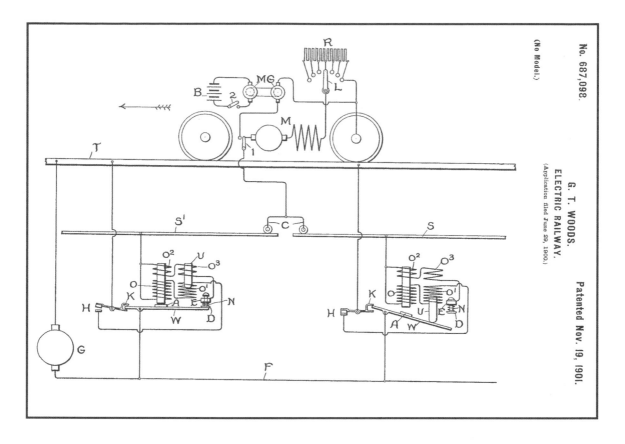

No. 687,098.

(No Model.)

G. T. WOODS.
ELECTRIC RAILWAY.
(Application filed June 29, 1900.)

Patented Nov. 19, 1901.

A diagram from one of Woods's numerous patents on electrical railway systems.

control of the business decisions, wanted only to sell his patents to other companies. Woods, who was supposed to receive a salary from Woods Electrical, was paid irregularly.

By 1890 Woods had extricated himself from his agreement with Woods Electrical and moved to New York City. Although the induction patents were controlled by Woods Electrical, Woods hoped to find different backers for some of his new ideas.

Investors were scarce and Woods again descended into poverty. In August 1891 Woods saw an advertisement for the American Patent Agency, which promised to help inventors sell their work. Intrigued, Woods visited American Patent, where he met James Zerbe, the manager. Zerbe impressed Woods, who entered into a partnership with him later that month.

Woods quickly regretted this decision when he discovered that Zerbe was filing, under Woods's name, patents that contained the ideas of other inventors. By October, Woods and Zerbe were getting into fistfights at the office. Woods broke off the partnership and testified before the Patent Office against the filings that Zerbe had made under his name.

Zerbe attempted to market an electric railway system that Woods had invented, building a demonstration rail at Coney Island in

Brooklyn. The railway, unveiled in February 1892, drew considerable attention because it was one of the first to rely on a third rail. The third rail, which forms a part of most modern subway systems, is a grounding rail that carries the electricity needed to power a train. Previous trolley systems had relied on overhead wires to carry electricity, but these wires were constantly exposed to wind and weather and, as a result, broke often.

Woods took out advertisements in newspapers informing the public that Zerbe had built the railroad using ideas he did not own. Zerbe retaliated by charging Woods with libel. The libel case went to trial in April 1892, and events began to favor Woods: he was quickly acquitted, with the jury granting him full control over his electric railway inventions.

Woods was not the only inventor Zerbe had treated badly; following Woods's successful defense, he and a group of other inventors began a campaign to have Zerbe barred from representing inventors at the Patent Office. (At the time, anyone could act as a patent solicitor who had not

Testing the electrical power of the third rail in the Metro subway system in Washington, D.C., in 2005.

been specifically forbidden to do so by the office.) In November 1893, they succeeded, and Zerbe was banned from patent practice.

LIFE AS AN INVENTOR

Ironically, Woods's problems with Zerbe probably benefited him in the long run. Despite its illicit origins, the Coney Island electrical railway built by Zerbe attracted considerable media coverage. Also, Woods's efforts to ban Zerbe from patent practice, in addition to his willingness to testify against his own patents to protect other inventors, helped to connect him to the inventing community. The libel trial put a spotlight on Woods's talents as an inventor. By 1896, the *New York Times* was referring to Woods as "one of the best known colored men" in the city.

By that point, Woods was able to attract the interest of legitimate technology companies that would actually develop his inventions and compensate him for his patents. Woods eventually developed very close relationships with General Electric Company and Westinghouse Electric and Manufacturing Company; he had more than forty patents to his name by the time of his death in January 1910 at the age of 54.

Woods continued to focus on railroad technology later in his life, particularly the electric trains that eventually became subway systems, an interest that dovetailed with General Electric's and Westinghouse's business strategies. In addition to the third rail, Woods helped to develop the automated air brake, which greatly improved train safety; and he made numerous other technological improvements to electric trains, including developing methods to better regulate their speed. His work helped to make train travel safer and contributed significantly to the development of modern urban subway systems.

—Mary Sisson

Further Reading

Books

Aaseng, Nathan. *Black Inventors*. New York: Facts On File, 1997.

Brodie, James Michael. *Created Equal: The Lives and Ideas of Black American Innovators*. New York: Morrow, 1993.

Fouché, Rayvon. *Black Inventors in the Age of Segregation: Granville T. Woods, Lewis H. Latimer, and Shelby J. Davidson*. Baltimore, MD: Johns Hopkins University Press, 2003.

Pursell, Carroll W., ed. *A Hammer in Their Hands: A Documentary History of Technology and the African American Experience*. Boston: MIT Press, 2005.

Web sites

Granville Woods

 A profile of Woods hosted by the Black Inventor Online Museum.
 http://www.blackinventor.com/pages/granvillewoods.html

The Multiplex Telegraph

 A profile of Woods from the Lemelson-MIT Program.
 http://web.mit.edu/INVENT/iow/woods.html

See also: Faraday, Michael; Stephenson, George; Transportation.

ORVILLE WRIGHT AND WILBUR WRIGHT

Inventors of the modern airplane

1871–1948 and 1867–1912

Humans have dreamed of flying since ancient times. Those dreams became reality in the early 20th century because of the work of two brothers, Orville Wright and Wilbur Wright, from Dayton, Ohio. The Wrights did not invent flying: others had already flown kites, balloons, gliders (unpowered airplanes), and model aircraft. Their genius was to bring together two great 19th-century inventions—the unpowered glider and the gasoline engine—to make the world's first powered and piloted airplane.

EARLY YEARS

Wilbur and Orville Wright were two of five children born to Milton Wright, a bishop of the Church of the United Brethren in Christ, and his wife Susan. Wilbur, the couple's third child, was born in 1867, on a farm outside Millville, Indiana; Orville was born four years later, after the family had moved to Dayton. Independent and businesslike from an early age, the two Wrights earned pocket money by selling homemade toys. A toy their father bought for them in 1878, a model helicopter powered by a rubber band, sowed a seed in their minds for an invention that would change history.

In 1886, at the age of 15, Orville Wright started his own school newspaper, *The Midget*, after his family gave him a printing press as a gift. Soon, the brothers had opened a print shop and built large presses of superb quality for other local printers. By 1889, Orville was publishing Dayton's weekly newspaper, *West Side News*, with his brother as editor, and charging 40 cents for a year's subscription. Their efforts to publish *The Evening Item*, a daily paper, were short-lived, largely because Dayton already had 12 other newspapers. Their next venture was more successful: their superb mechanical skills led them to open a bicycle shop in December 1892. The Wright Cycle Company had its own workshop and by 1896 was manufacturing bicycles for sale or rent. The brothers were known not only for their skills but also for their appearance, wearing suits and bow ties even when fixing tires or straightening spokes.

Orville Wright, left, and Wilbur Wright, both photographed in 1905.

STUDYING FLIGHT

On August 10, 1896, the brothers and the world were shocked by the death of German engineer Otto Lilienthal (1848–1896), who died after crashing one of his experimental gliders. Often described as a "birdman," Lilienthal helped to prove that humans could fly by strapping enormous wooden wings to his body and jumping from a hill. The Wrights had taken a close interest in flight ever since their father bought them a model of an early helicopter almost twenty years before. Soon after hearing of Lilienthal's death, they turned their attention to making a flying machine of their own.

The brothers started off by poring over books in the library, reading about all those who had tried to fly before them. They read about pioneers like the Montgolfiers, two French brothers who flew the world's first hot-air balloon in 1783. Hot-air balloons are difficult to steer, however. Gliders, which became popular during the second half of the 19th century, interested the Wrights more because they were easier to control.

During 1899, the Wrights read all the material they could find on flight and started writing to people who they thought might help, including Samuel P. Langley (1834–1906), head of the Smithsonian Institution, who had already flown large model airplanes. That summer, they built a two-foot (.6-m) kite with wings to try out their ideas. Dayton was flat and mostly windless, so by November they were writing to the U.S. Weather Bureau asking where the best windy places were to test gliders. Their most important series of letters was begun on May 13, 1900, when Wilbur wrote to Octave Chanute (1832–1910), a world-famous engineer who helped early aviators. Chanute offered to meet the Wright brothers and exchanged several hundred letters with them over the next 10 years.

> The person who merely watches the flight of a bird gathers the impression that the bird has nothing to think of but the flapping of its wings. As a matter of fact this is a very small part of its mental labor. . . . We only learn to appreciate it when we try to imitate it.
>
> —Wilbur Wright

TAKING TO THE AIR

Encouraged by Chanute's enthusiasm, the Wrights made their first serious experiments from the windy Kill Devil Hill, near the village of Kitty Hawk in North Carolina. Their first glider was a 10-foot (3-m) biplane with two wings, one stacked above the other, but they flew it mostly as a kite, without a pilot, to see how it behaved. A glider has no engine, so it needs big wings to generate enough lift (upward force) to support its own weight plus the weight of the pilot. The Wrights' first glider had wings that were too small, so it flew no more than 300 feet (91 m) before crashing to the ground. The following year, in 1901, they made a

Who Flew First?

On May 6, 1896, seven years before the Wright brothers made their historic flight, Samuel Pierpont Langley
flew a 16-foot (5-m) model airplane for about ninety seconds
across the Potomac River. Langley's extraordinary plane was
powered by a steam engine and launched from a riverboat by a cat-
apult, but it could not carry a pilot. The years that followed the
Wrights' success saw some debate about whether Langley had really
invented the "flying machine" first. That question became a source of
great annoyance to the Wrights.

Langley, who was the secretary of the Smithsonian Institution
and one of the most distinguished scientists in the United States,
could not have been more different from the Wright brothers, the
two unknown bicycle makers from Dayton, Ohio. He enjoyed
government backing for his plane, but the Wrights had to struggle to finance their own experiments from the profits of their
bicycle shop.

However, whereas the Wrights had achieved glory, Langley's
efforts had ended in bitter disappointment. Despite having $50,000
from the War Department to develop person-carrying airplanes, and
a similar sum from the Smithsonian, Langley's failures became
something of a joke. One government official told the newspapers:
"You tell Langley for me that the only thing he ever made fly was
government money."

After Langley's death in 1906, the leaders of the Smithsonian
remained fiercely loyal to him: they insisted he had built a machine
"capable" of flying years before the Wright brothers and deserved credit.

Orville Wright considered this an insult and demanded an apology. None was forthcoming, so instead of lending *Flyer* to the
Smithsonian, as many people would have expected, he had it shipped
to the Science Museum in London, England. It stayed there until
1942, when the Smithsonian Institution issued an apology. The plane
was promptly shipped back to the United States and to the
Smithsonian, where it is a permanent part of the collection at the
National Air and Space Museum.

The Wright brothers' third test glider, launched on October 10, 1902, at Kill Devil Hills, North Carolina. Orville Wright is at left; Wilbur Wright is aboard the glider; their friend Dan Tate is to the right.

glider with wings almost twice the size of the original, but, to the Wrights' great disappointment, it flew only about 100 feet (30 m) farther.

That fall, they discovered the problem. The brothers were designing their gliders using numeric data that Langley and Lilienthal had worked out—and the data seemed to be wrong. Having decided that they needed better data, they built a wind tunnel (an enclosed chamber with a fan at one end) out of a wooden box, a gasoline engine, and an old fan. With a measuring device made from an old hacksaw blade and some bicycle spokes, they tested more than two hundred different wing shapes inside the wind tunnel until they hit upon the best design. With a new glider in 1902, they found they could now fly more than 600 feet (183 m) and stay airborne for almost 30 seconds. Although much trial and error was involved, the Wrights were extremely scientific in their approach. They made almost one thousand flights, taking photographs of their tests and recording all the details in notebooks.

The Wrights' glider still relied on wind power. During late 1902 and 1903, the brothers started to wonder how they could make a powered airplane that did not need any wind. The obvious answer was to use a gasoline engine like the ones that had recently been invented for automobiles. Gas engines were huge and heavy; how could an airplane carry one and still get off the ground? When the Wrights asked engineering companies to help them build a lightweight engine, no one knew how. In the true spirit of invention, the brothers returned to their bicycle workshop in Dayton and built their own engine. By September 1903,

How Wing Warping Works

". . .I have been afflicted with the belief that flight is possible to man. . . ."

In Wilbur Wright's remarkable letter to engineer Octave Chanute, he declared: "For some years I have been afflicted with the belief that flight is possible to man. My disease has increased in severity and I feel that it will soon cost me an increased amount of money if not my life." After years of study and experimentation, the Wright brothers' "disease" bore fruit when *Flyer* took to the air on December 17, 1903. Thinking of their plane as a glider, the Wrights had chosen the windiest place they could find, Kitty Hawk in North Carolina, with steep dunes to help them take off and soft sand to cushion their landing. In fact, their biggest achievement was to prove that airplanes could fly without any wind at all.

Gliders stay in the air because they have airfoil wings: seen from the side, the top of an airfoil is curved and the bottom is flat. Air has to move farther over the top of the wing than under the bottom, so the air that travels over the top of the wing has to go faster to keep up. This fast-moving air creates an upward force (lift) that keeps a glider airborne. A glider relies on wind to blow over the wings and create lift. The Wright brothers realized they could achieve the same effect by putting an engine and two propellers at the back of their plane so that, in effect, it made its own wind. The propellers created a back-draft of air that pushed the airplane forward. Its wings created lift and pushed it up into the air.

Becoming airborne was one thing, but staying aloft was quite another. In his letter to Octave Chanute of May 1900, Wilbur

The first heavier-than-air flight, on December 17, 1903, with Orville Wright piloting and Wilbur Wright alongside the plane.

Wright explained how he had watched buzzards steering themselves and controlling their flight by twisting the tips of their wings. This was the inspiration for *Flyer*'s steering system, which the Wrights called "wing warping," and which used cables running from the pilot's waist to the wings. To steer the plane, the pilot wriggled one way or the other so the cables pulled the back edge of one wing up and the other wing down. This made more lift on one side of the plane than on the other and steered the plane to the left or to the right. It seemed a strange and frightening system, but it worked.

An airplane cannot reach speed and become airborne immediately, so the Wright brothers still needed a means to launch *Flyer* into the air. They achieved launch by mounting the plane on a truck that ran along a short length of wooden railroad, which they had previously laid down on the sand dunes at Kill Devil Hill. With the engine buzzing and the propeller turning, they let the plane roll down the track until it was going just fast enough to lift itself in the air. Wilbur, running alongside, steadied one wingtip with his hands. Strapped in the pilot's position and facing forward, Orville flew into the future—and straight into history.

How Wing Warping Works

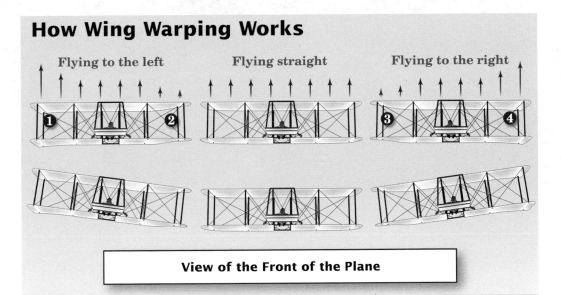

Flying to the left

Flying straight

Flying to the right

View of the Front of the Plane

Suppose the pilot tilts the back of the right wing ❶ down and the back of the left wing ❷ up. There is now more lift on the right wing, so that the right wing rises, the left wing falls, and the plane banks to the left.

With the back wing flaps straight, the wings produce equal lift across their surface and the plane flies in a straight line.

Now the pilot tilts the back of the left wing ❸ down and the back of the right wing ❹ up. There is now more lift on the left wing, so that the left wing rises, the right wing falls, and the plane banks to the right.

their new plane, *Flyer* or *Kitty Hawk Flyer*, was ready (the brothers themselves often called it "the Whopper flying machine"). They returned to Kitty Hawk and spent several weeks preparing to fly. Finally, on December 17 at 10:35 a.m., Orville Wright took to the air. In the 12 brief seconds that he stayed aloft, he made history as the first person ever to fly an engine-powered airplane (see box, How Wing Warping Works).

THE PRACTICAL PLANE

The Wright brothers' first powered flight was an important scientific experiment as well as a historic achievement: it proved that a powered airplane could carry a person through the sky. However, *Flyer* traveled less than 300 yards (274 m), a distance that was not much of an improvement over their gliders. The Wright brothers knew they would have to fly much farther, and they set about making an improved version of *Flyer*. By the following summer, they were flying almost three miles (5 km) for up to five minutes at a time. Hardly anyone knew what they were doing; the Wrights decided to keep their work secret until they had protected their invention. Meanwhile, they improved the plane's design and made more trials.

In 1905, they twice offered their invention to the U.S. government— and twice they were turned down. By this time, the plane could fly 24.5 miles (39.4 km) in 39 minutes at an average speed of 38 miles per hour (61 km/h). In May 1906, a patent was granted to the Wright brothers for their "Flying Machine." More confident now about showing their invention, they began making demonstration flights in the United States and Europe.

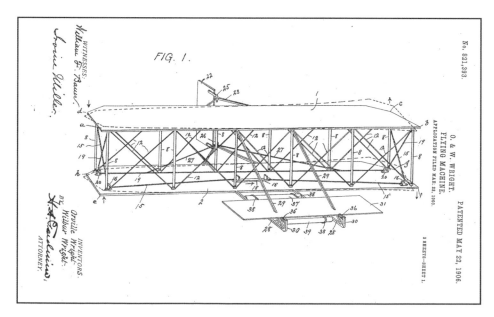

A perspective view of the Wright brothers' flying machine from their 1906 patent.

The year 1908 brought great success when the U.S. War Department finally accepted the Wrights' offer to build planes, but disaster struck, too. During a test flight on September 17, Orville Wright's airplane crashed to the ground from a height of 75 feet (23 m), severely injuring him and killing his passenger, Lieutenant Thomas Selfridge—the first person to die in an airplane accident.

The following year, 1909, the Wrights set up an airplane manufacturing company in New York City, with Wilbur as president and Orville as vice president. They also allowed some industrialists in Germany to set up a company, Wright GmbH, to make their planes under license in return for a large payment. Almost from the start, much of their energy was taken up fighting rivals who seemed to be copying their ideas. Their first target was Glenn Curtiss (1878–1930), soon to become a famous pilot, who started to manufacture his own airplane in June 1909. Other patent suits followed in both the United States and Europe, though they did not really solve the problem. The legal wrangling over their invention finally ended in 1917 when their patent expired; the U.S. government refused to renew it, because competition would lead to better airplanes.

The Wright brothers, photographed at the International Aviation Tournament in 1910.

LATER YEARS

Legal battles and other business worries took their toll on Wilbur Wright. Suffering from typhoid fever, he died on May 30, 1912, at the age of 45.

For the next three years, Orville ran the Wright Company alone, but in 1915 he sold his stake and retired from the airplane-making business. From then on, he led a quieter life, tinkering with machines and inventions, collecting awards, promoting aviation, and spending time with his friends and family in Dayton and at a country retreat in Georgian Bay near Lake Huron in Ontario, Canada. Orville died of a heart attack in Dayton on January 30, 1948.

AIRPLANES AFTER THE WRIGHTS

During the 20th century, airplanes revolutionized around-the-world transportation as much as railroads had changed travel over land a century earlier. Pilots were soon flying farther, faster, and for longer—crossing oceans, setting records, and effectively making the world smaller. World War I (1914–1918) greatly speeded the development of better airplanes, as rival nations raced to gain advantage in the air.

World War II (1939–1945) produced a dramatic leap forward when British engineer Frank Whittle (1907–1996) and his German rival Hans Pabst von Ohain (1911–1998) produced the jet engine that was soon flying airplanes at the speed of sound (around 660 mph, or 1,062 km/h). The jet

Flyer on display at the National Air and Space Museum in Washington, D.C.

TIME LINE

1867	1871	1892	1896	1900	1903
Wilbur Wright born near Millville, Indiana.	Orville Wright born in Dayton, Ohio.	The Wright Cycle Company is founded.	The Wright brothers begin studying flight.	The Wright brothers build and crash their first glider.	The Wright brothers build *Flyer* and successfully test it.

engine arrived too late to make much impact on the war itself, but it certainly changed travel in the decades that followed: from the 1950s onward, jet airplanes led to a new era of inexpensive, international air travel. At the same time, rocket engineers were sending the first satellites and people into space. All these developments built on the Wright brothers' accomplishments.

Many people think of the Wrights as fresh-faced country boys who stumbled on the secret of flying by accident. In fact, they researched their ideas for many years, meticulously tested their airplanes, built their machines with immense technical skill, and protected their ideas like shrewd businessmen. They knew their invention was valuable: they could see that airplanes would eventually carry people and cargo around the world—they even believed their planes might stop wars. Their confidence in the importance of their invention led them to perform tests in secret and to sue rivals. They believed their patent meant that they alone owned the right to make powered airplanes, but they had benefited from earlier inventions and help from others, including Otto Lilienthal and Octave Chanute, and some people considered them and their actions small-minded and selfish.

After Wilbur's death and Orville's decision to sell his patent interests in 1915, the Wright brothers pretty much disappeared from the aircraft business. In the 21st century, Ford cars, McCormick harvesters, and Goodyear tires still recall the names of their original inventors, but no Wright airplanes crisscross the skies. Nonetheless, the part played in airline history by the two ingenious brothers has never been forgotten.

—Chris Woodford

The U.S. Navy's Blue Angels precision flying team practicing maneuvers in 2003.

TIME LINE (continued)

1906	1909	1912	1915	1948
The Wright brothers patent their flying machine.	The Wright Brothers found an airplane manufacturing company.	Wilbur Wright dies.	Orville Wright retires.	Orville Wright dies.

Further Reading

Books

Carson, Mary. *The Wright Brothers for Kids: How They Invented the Airplane*. Chicago: Chicago Review Press, 2003.

Freedman, Russell. *The Wright Brothers: How They Invented the Airplane*. New York: Holiday House, 1994.

Web sites

Reliving the Wright Way
   Information, games, lesson plans, and other activities developed by NASA.
   http://wright.nasa.gov/

Wilbur and Orville Wright Papers at the Library of Congress
   A collection of photographs, letters, and other materials.
   http://lcweb2.loc.gov/ammem/wrighthtml/wrighthome.html

Wright Flyer Project
   A project to build a modern replica of the Wright brothers' plane.
   http://www.wrightflyer.org/

See also: Diesel, Rudolf; Sikorsky, Igor; Transportation; Whittle, Frank; Zeppelin, Ferdinand.

YOUNG INVENTORS

In 1994, four-year-old Sydney Dittman of Houston, Texas, became the youngest person ever to be awarded a patent—for a device that helps disabled people grasp doorknobs. Although hers was an exceptional achievement, many individuals show signs of creativity at an early age and begin inventing in their youth. Some of the world's greatest inventors, including Thomas Edison, the Lumière brothers, and Benjamin Franklin, began inventing when they were still only teenagers.

PROLIFIC AND PRECOCIOUS

That prolific inventors (those who devise many different inventions) often start young should not be surprising. Benjamin Franklin (1706–1790) was one of the best-known 18th-century Americans. Apart from being a Founding Father of the United States, he was also a journalist, printer, scientist, and tireless inventor. He developed his first invention, a set of fins to help him swim more quickly, around the age of 10, and invented many other devices later, including the lightning rod.

Illustrations from Sydney Dittman's patent for a "tool for grasping round knobs"; at age four, Dittman became the youngest inventor to receive a U.S. patent.

Like Benjamin Franklin, British inventor Henry Bessemer (1813–1898) began early. As a boy, he became fascinated with casting (making objects out of molten metal). This led indirectly to his most famous invention: a method of making steel in large quantities that played an important part in the construction industry from the 19th century onward. He made his first invention at age 17, when he developed a new kind of stamp that could not be counterfeited.

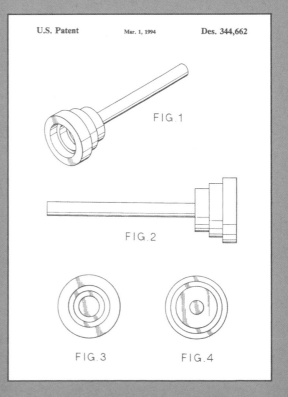

U.S. Patent Mar. 1, 1994 Des. 344,662

FIG.1

FIG.2

FIG.3 FIG.4

Another prolific British inventor, Charles Wheatstone (1802–1875), began making unusual musical instruments at the age of 16. That led him to experimental studies of sound, which established him as one of the greatest scientists of his day. A more recent musical pioneer is Les Paul (1915–), inventor of the solid-bodied electric guitar. At the age of seven, he was tinkering with his mother's automated player piano; two years later, he built himself an electronic radio. His numerous contributions to musical technology have earned him the nickname "the Edison of pop."

Thomas Edison (1847–1931) was perhaps the most prolific inventor of all time. Eager to establish his independence, he worked on the railroads in his mid-teens and by the age of 15 was writing and printing his own newspaper. Over the next few years, he learned to operate a telegraph and improved its mechanism in various ways. Before the age of 20, he had invented a new piece of telegraph equipment—the repeating telegraph—that could relay messages down the line, through unmanned offices, but his ingenuity earned him nothing more than a reprimand from his boss for wasting company time.

PASSION FOR TECHNOLOGY

Like Edison, many young inventors start from a fascination with the latest technology. As a boy, Tim Berners-Lee (1955–), British-born inventor of the World Wide Web, built a make-believe computer out of old cardboard boxes and sticky tape. Within a few years, he had graduated to making electronic circuits to control his model railroad. The young Steve Wozniak (1950–) enjoyed reading science fiction novels about Tom Swift, a teenage superhero who used science to save the world. Later, Wozniak became an electronics buff, then one of the inventors of the Apple personal computer in the 1970s. Since leaving Apple, he has spent much of his time teaching electronics to elementary school children in California in the hope that some will become tomorrow's great inventors.

Technology of a different kind inspired an earlier computer pioneer, Blaise Pascal (1523–1662). When he was only 19, he built one of the world's first calculators using wooden gears to add and subtract numbers. This was the only major invention of Pascal's career, but he went on to become a great mathematician and philosopher. Not

One famous young inventor is Thomas Edison, photographed here at age 14.

all computer pioneers have such brilliant beginnings. Linus Torvalds (1969–), inventor of the Linux operating system, claims to have started out as "a nerd—before being a nerd was considered to be a good thing." He became famous in his twenties.

YOUNG ENGINEERS

Tinkering with machines is often a point of entry for young inventors. Some are inspired by watching parents or older friends experimenting and decide to try inventing themselves. Isambard Kingdom Brunel (1806–1859) became a great engineer after working with his father, who was also a famous inventor. Steve Wozniak learned from his father, an aerospace engineer. Steve Jobs (1955–), who cofounded Apple Computer with Wozniak, befriended an older neighbor who made electronic appliances from kits.

Sometimes young engineers make their way by resolving to do better than their parents. As a teenager, Cyrus Hall McCormick (1809–1884) watched his farmer father struggling to automate the harvest by inventing a machine that could cut the crops in the fields. When

Driven by Dreams

Some of the world's greatest inventors were spurred on by dreams they had in childhood. The father of the modern space rocket, Robert H. Goddard (1882–1945), first had the idea of traveling to the stars at age 17 when he climbed a cherry tree in his parents' garden and stared up at the sky. The 12-year-old Henry Ford became interested in making vehicles after seeing a coal-powered steam tractor on his father's farm. That event, which stayed with him all his life, prompted him to invent cars and tractors that changed the world.

The brothers Orville Wright and Wilbur Wright (1871–1948; 1867–1912) also made history with a childhood fascination. When the Wright brothers were boys, their father presented them with a model helicopter powered by a rubber band—a toy that led them to invent the modern airplane.

For some young inventors, inspiration comes with dreams of rebellion. As a boy, Indian physicist Narinder Kapany (1927–) was taught that light always travels in straight lines, so he immediately became determined to make it curve around corners. This obsession led him to pioneer fiber optics (a way of bending light down cables to carry images or telephone calls) in the 1950s.

his father gave up on the invention, Cyrus took over and eventually made it work. From this simple beginning, he founded one of the world's greatest agricultural machinery corporations. Henry Ford (1863–1947) was another inventor who formed a famous company after growing up on a farm. As a boy, he found the farm chores tedious and began tinkering with machines that could automate them. This started him on the path that led to the Ford Motor Company, which mass-produced affordable automobiles and farm tractors.

Eli Whitney (1765–1825) was an earlier pioneer of mass production and another precocious engineer. At the age of eight, he repaired his father's watch; four years later, he was building violins. A few years later, he set up a small business making nails and hat pins, which helped to fund his studies at Yale. Soon afterward, his invention of the cotton gin helped to start an agricultural revolution in the United States.

MAKING LIFE BETTER

For some young people, invention is a way to overcome adversity. As a teenager, photocopier pioneer Chester Carlson (1906–1968) was determined to become an inventor so he could escape the pover-

Older Inventors

Youth can be an advantage to inventors, but age and experience are usually helpful, too. Popular British inventor Trevor Baylis (1937–) had been inventing for some years when he finally won international acclaim in 1996, at age 59. The idea that made him famous was an inexpensive wind-up radio for people in developing countries. The inspiration for this invention was a television program that had explained that improved communication in poorer nations would raise awareness about issues like HIV/AIDS and save many lives. Baylis was still a "young inventor" compared with the oldest person ever to have been awarded a patent: American physicist Charles Greeley Abbot (1872–1973), whose final patent was awarded the year he died at age 101.

Not surprisingly, some famous older inventors started life as famous young inventors. Thomas Edison filed his last patent just months before he died, at age 84. Alexander Graham Bell also was inventing till practically the day he died. He received his last patent, for a hydrofoil boat, shortly before his death at age 75.

ty he had known in childhood. French inventor Louis Braille (1809–1852), who became blind in an accident at age three, escaped a different kind of hardship. When he was 15, he developed a writing system based on raised dots that enabled him and his friends to read books with their fingertips. Braille invented out of personal necessity, but he made life better for blind people everywhere.

The telephone pioneer Alexander Graham Bell (1847–1922) was not disabled, but his mother was deaf and Bell spent his early career as a teacher of deaf children. His lifelong interest in helping the deaf led to various inventions connected with speaking and hearing, and ultimately to his invention of the telephone in 1876. Bell was a prolific inventor who made his first invention (a device for cleaning wheat) when he was only 11.

Deafness also afflicted the early life of Thomas Edison. Before his hearing difficulties were identified, teachers thought he was suffering from a learning disability, which prompted his parents to school him at home. Fortunately, this instilled in him a curiosity and love of learning that played an enormous part in his career as an inventor.

Helping those less fortunate has remained a source of inspiration for young inventors. African American doctor Patricia Bath (1942–) made her name in the 1980s when she invented a new type of laser surgery for people with cataracts (an eye disorder that can lead to blindness). However, her brilliance and concern for others had been evident for years beforehand. At age 16, she carried out advanced research on cancer, and she contributed to an international scientific paper only one year later.

YOUNG INVENTORS TODAY

Young people of the 21st century still grow up tinkering with machines and electronics; but in a more technologically advanced age dominated by huge corporations, it is harder for them to become the next Edison, Ford, or Bell. For two young men, Steve Wozniak and Steve Jobs in the 1970s, putting together their rudimentary Apple personal computer in a garage was comparatively easy and it made their fortunes. Just a few decades later, many millions of dollars of investment would be needed to produce a computer better than those already available in stores. In the same way, young inventors have little hope of developing a revolutionary automobile or a better kind of electric power. The reality of modern inventing is that most creative young people will go to work for large corporations, such as IBM or DuPont, where their ideas will be pooled with those of others.

Does this mean that society has no place for young inventors anymore? Quite the contrary. New technology and improved ideas will always be needed, and young people are sometimes better placed to supply them; young people frequently have greater drive and enthusiasm and may be more in touch with the latest technology. Young people also tend to be more idealistic and optimistic than their elders; these qualities are always an asset for inventors. From Benjamin Franklin to Linus Torvalds, young inventors have typically used their talents to change the world for the better. Society will always need them.

—Chris Woodford

Further Reading

Books

Bridgman, Roger. *1,000 Inventions and Discoveries*. New York: Dorling Kindersley, 2006.

Caney, Stephen. *Stephen Caney's Invention Book*. New York: Workman, 1985.

Tucker, Tom, and Richard Loehle. *Brainstorm! The Stories of Twenty American Kid Inventors*. New York: Farrar, Straus, and Giroux, 1998.

Web sites

Great Sites for Young Inventors.
How to get involved in inventing.
http://inventors.about.com/od/kidactivities/

Young Inventors International
A nonprofit group dedicated to helping budding inventors.
http://www.younginventorsinternational.com/

See also: Bath, Patricia; Bell, Alexander Graham; Berners-Lee, Tim; Bessemer, Henry; Braille, Louis; Brunel, Isambard Kingdom; Carlson, Chester; Edison, Thomas; Ford, Henry; Franklin, Benjamin; Goddard, Robert H.; Jobs, Steve, and Steve Wozniak; Kapany, Narinder; McCormick, Cyrus Hall; Paul, Les; Torvalds, Linus; Wheatstone, Charles; Whitney, Eli; Wright, Orville, and Wilbur Wright.

FERDINAND VON ZEPPELIN

Inventor of the rigid dirigible

1838–1917

Ferdinand von Zeppelin was a member of the German aristocracy and enjoyed some celebrity long before he dedicated his life to creating a machine that could be guided through the air. Although Zeppelin faced many obstacles, he created a pioneering form of air transportation, becoming as famous for his determination as for his invention.

EARLY YEARS

Ferdinand von Zeppelin was born in 1838 on a family estate on Lake Constance in the kingdom of Württemberg, located in what is now southern Germany. Zeppelin's father was a count, but his parents did not like the ceremony of court life and had moved to the country. As a result, Zeppelin had a somewhat unusual upbringing for a member of his class. He was taught to work with his hands, and he associated with people from many walks of life.

His position as a young German count, however, obligated him to become an officer in the army. Zeppelin attended the War Academy at Ludwigsburg; at graduation in 1857, he became a lieutenant. Württemberg was at peace, however, and Zeppelin quickly lost interest in military service. The next year he received permission to attend the University of Tübingen to study engineering. After a year at Tübingen, he rejoined the army as a member of the engineering corps.

ADVENTURER

Zeppelin was still a young man and he wanted adventure. In the United States, the Civil War had broken out; in 1863, Zeppelin traveled there as an observer attached to the Union Army.

While in the United States, Zeppelin made his first ascent in a lighter-than-air vehicle, a hot-air balloon. At the time, balloons were a relatively new invention and they fascinated Zeppelin. He thought they might be useful in battle: a balloonist could float over enemy lines and observe enemy actions or drop bombs.

Zeppelin returned to Württemberg. In 1870, Prussia, which was allied to Württemberg, went to war with France. Zeppelin led a daring raid into French territory to gather intelligence; of the 11 men in the party, only

A portrait of Ferdinand von Zeppelin from 1907.

Zeppelin survived, bringing important news about French troop movements to the Prussians.

During this period, Zeppelin observed the uses of balloons in war: during a siege of Paris, the French evaded Prussian troops by using balloons to carry messages and people. Although the balloons were impressive, they were limited: they could not be steered, and if the winds were unfavorable, they would crash. Zeppelin began to wonder about building a balloon that could be piloted—a dirigible.

A NEW OCCUPATION

In 1890, Zeppelin was forced to retire from the army for reasons now unknown. He immediately threw himself into dirigible research, hiring an engineer and drawing up plans in hopes that the Prussian military would fund construction.

Zeppelin's idea was for a lighter-than-air craft that could be controlled by a pilot. He wanted the dirigible to be able to travel long distances and carry heavy loads, so he decided to make it rigid. Unlike balloons or blimps, Zeppelin's dirigibles had metal skeletons that gave them their distinctive cigar shape—a shape chosen to make the vessels easier to steer. To increase safety, each dirigible contained many balloons, rather than just one.

The LZ-1 airship, piloted by Zeppelin, on its maiden flight over Lake Constance in 1900.

The Prussian government, however, was unwilling to pay to build the experimental craft. Zeppelin then submitted his plans to a national association of engineers, which supported them and put forward an appeal for private funds. Zeppelin used the money raised, as well as his own money, to build his first dirigible, the *LZ-1* (LZ is an abbreviation for *Luftschiff Zeppelin*, a German phrase meaning "Zeppelin airship").

CRASHES AND RECOVERIES

The *LZ-1* made its first flight over Lake Constance in July 1900. The 420-foot (128-m) long dirigible flew for 17 minutes and landed safely; in October it made two more short flights. The flights were not complete successes, however. The dirigible had engine troubles and at times it just drifted in the wind, seeming little better than a regular balloon.

Zeppelin wanted to build an improved dirigible, but as the *LZ-1's* flights were unimpressive, the Prussian government continued to decline his requests for funding. Zeppelin finally sent out mass mailings and took out advertisements in papers asking people for money; in addition, the king of Württemberg held a fund-raiser. That money, combined with the last of Zeppelin's personal fortune, was enough to build the *LZ-2*.

> It was never necessary for me to fight for self-confidence and faith for the simple reason that mathematics, logic, and practical experiments gave me one proof after another that I was right.
>
> —Ferdinand von Zeppelin

The *LZ-2* made its first and last flight in January 1906. It was an improved design, with a more rigid frame and stronger engines, but the engines died and Zeppelin had to make an emergency landing, tying the ship down in a field. That night, a violent storm arose, and the *LZ-2* was destroyed.

Zeppelin again asked the Prussian government for funds and was again denied, but the king of Württemberg held another fund-raiser. In October 1906, the *LZ-3* made two test flights. By this point, Zeppelin had significantly improved the design, using fins along the dirigible's sides and tail to keep it stable. Better engines powered the ship through the air at more than 30 miles per hour (48 km/h). The *LZ-3* was clearly what Zeppelin had promised—a vehicle that could be driven through the air, rather than one that floated passively. Two years later, Zeppelin developed the *LZ-4*, which made lengthy flights and convinced the Prussian government—and the German public—of the usefulness of the rigid dirigible.

THE LIMITS OF THE ZEPPELIN

Zeppelin then began manufacturing dirigibles in earnest; by 1909, he had built enough to start offering commercial air travel in Germany. When World War I broke out in 1914, the Prussian military ordered dozens of Zeppelin's dirigibles to use for bombing and reconnaissance operations.

The dirigibles did not fare well in the war, however. They floated because of hydrogen, a flammable, lighter-than-air gas; but the hydrogen, combined with the dirigibles' slow speeds, made them vulnerable to artillery fire.

In addition, dirigibles were outfought by a newer aviation technology: the airplane. Although dirigibles could fly higher and longer than the airplanes of that era, being forced to fly at very high altitudes made

Catching Fire

After the flight of the *LZ-3* in 1906, the Prussian government gave Zeppelin money to build one more dirigible, promising more money if he could successfully fly the new dirigible for 24 hours nonstop. Zeppelin built the *LZ-4*, taking it on short flights in June 1908 and then on a 12-hour tour over Switzerland in July.

In August 1908, Zeppelin launched the *LZ-4* on what he hoped would be the vital 24-hour flight. Mechanical troubles forced an early landing, however, and once the *LZ-4* was on the ground, the wind blew it out of control. A fire started, and the hydrogen in the dirigible ignited, destroying it and killing some of Zeppelin's workers.

What seemed like a serious blow to Zeppelin's invention wound up proving how much the dirigible had caught the imagination of ordinary Germans. At a time when any airship was rare, the accomplishments of the rigid dirigible—which was already being called a zeppelin—had become a source of great pride. Zeppelin's determination, especially in the face of discouragement from the Prussian government (which was resented by many Germans), also was inspiring. Already, Germans were eagerly buying zeppelin souvenirs and beginning to work the inventor's name into songs and poems—the beginnings of what would be the full-fledged craze of zeppelinism.

Within a day of the *LZ-4* disaster, German towns and private citizens were sending Zeppelin money to build another dirigible. Eventually so much money was sent to Zeppelin that he was able to create a permanent foundation to fund the development of his dirigible. Public opinion forced the Prussian government to become more generous as well, and it repaid Zeppelin all the money he had spent out of his own pocket to develop his dirigible.

TIME LINE

1838	**1857**	**1863**	**1890**	**1900**
Ferdinand von Zeppelin born in the kingdom of Württemberg.	Zeppelin graduates from the War Academy at Ludwigsburg.	Zeppelin travels to the United States as an observer for the Union Army during the Civil War.	Zeppelin retires from the army and begins dirigible research.	Zeppelin's dirigible, the *LZ-1*, makes its first flight over Lake Constance.

Illustration, published in 1920, of a German zeppelin flying over London during bombing raids of September 2–3, 1916.

aiming bombs or collecting intelligence harder. Airplanes were being improved upon as well, and they began using bullets designed specifically for use against dirigibles; these projectiles proved to be lethal. As the war continued, even Zeppelin came to the conclusion that his dirigibles were ill-suited to combat.

In February 1917, Zeppelin became ill with appendicitis and underwent surgery. While recovering from the operation, he contracted pneumonia; he died on March 8.

In November 1918, Germany conceded defeat and was forced to surrender its remaining zeppelins. Zeppelin's former colleagues rebuilt the commercial dirigible business, however, and in 1929 the dirigible *Graf Zeppelin* (Count Zeppelin) flew around the world. During the 1930s, zeppelins were used to make regular commercial flights from Germany to the United States.

In 1937, however, the zeppelin *Hindenburg* caught fire while landing in Lakehurst, New Jersey, and the explosion killed 36 people. The disaster, combined with the zeppelin's association in the public's mind with what was now Nazi Germany, permanently damaged its commercial appeal. Additionally, airplanes became larger, more comfortable, and more reliable. In the post–World War II era, airplanes became the transport of choice across the Atlantic Ocean.

In modern times, zeppelins are rare, and rides are offered only as a curiosity. Even the ready availability of helium, a nonflammable, lighter-than-air gas that is used in

TIME LINE

1906	1908	1909	1914	1917
The *LZ-2* and *LZ-3* are tested.	Zeppelin develops the *LZ-4*.	Zeppelin begins manufacturing dirigibles.	Zeppelin's dirigibles are used for bombing and reconnaissance operations during World War I.	Zeppelin dies of pneumonia.

The Hindenburg *ignites while trying to land in Lakehurst, New Jersey, in 1937, killing 36 people.*

balloons and blimps, has not brought back the zeppelin in any significant way. Nonetheless, Zeppelin's invention heralded the modern airplane industry, proving that flying machines could be practical and useful.

—Mary Sisson

Further Reading

Books

de Syon, Guillaume. *Zeppelin! Germany and the Airship, 1900–1939.* Baltimore, MD: Johns Hopkins University Press, 2002.

Goldsmith, Margaret. *Zeppelin: A Biography.* New York: Morrow, 1931.

Vaeth, J. Gordon. *Graf Zeppelin: The Adventures of an Aerial Globetrotter.* New York: Harper, 1958.

Web sites

The Hindenburg
An exhibit from PBS's show *Secrets of the Dead* on the fatal disaster.
http://www.pbs.org/wnet/secrets/html/e3-menu.html

Zeppelin Museum Friedrichshafen
Web site of the German museum dedicated to the rigid dirigible (in English).
http://www.zeppelin-museum.de/firstpage.en.htm

See also: Military and Weaponry; Transportation; Wright, Orville, and Wilbur Wright.

GLOSSARY

abacus The first, and simplest, calculator, consisting of beads (representing numbers) that slide along wires; enables people to carry out complex calculations.

aerodynamics Science of how air flows around moving objects.

aerospace Science concerned with the earth's atmosphere and beyond.

Age of Exploration Decades of the fifteenth and sixteenth centuries, marked by improvements in shipbuilding and the discovery of ocean routes from Europe to Asia and the Americas.

agriculturalist Scientist who studies crop and livestock production and soil cultivation; also, an individual who farms and practices animal husbandry.

algorithm Established and ordered set of procedures designed to arrive at a solution or solve a problem.

alloy Substance consisting of a metal mixed with another substance, either metal or nonmetal.

alternating current Electrical current flowing first in one direction, then in the other.

anaerobic Referring to organisms or systems that do not need oxygen to process and release the energy found in their food.

anatomy Scientific study concerned with major internal bodily structures.

anesthesia Treatment used to dull pain during surgery, administered in two forms: local (a limited area of the body is deadened to sensation) and general (the patient is made unconscious).

appertization Method of preserving food invented by Nicolas-François Appert, involving the canning, heating, and sealing of foods.

aqueduct Man-made channel built to carry large amounts of water from one point to another.

Archimedes' Principle Principle stating that any object immersed in a fluid will be subject to an upthrust (or buoyant force) equal to the weight of fluid that the object displaces.

arithmometer Popular term in the late 19th century for what is now called a calculator.

arms race Competition between the Soviet Union and the United States for superiority in nuclear weapons during the Cold War (late 1940s to the 1980s).

assembly line Production method involving the manufacture of goods by moving them down a line of workers; closely associated with mass production.

astronomy Scientific study concerned with objects and matter outside the earth's atmosphere.

astrophysics Branch of astronomy that studies the properties and behavior of planets, stars, and other celestial objects.

atom Smallest unit of a chemical element that can participate, alone or in combination, in a chemical reaction; atoms consist of smaller particles, including electrons, protons, and neutrons.

atomic energy Energy produced by splitting the nucleus of an atom.

bacteria (singular, bacterium) Unicellular organisms that usually exist in large number in colonies; found in water, organic matter, and the bodies of plants and animals.

Bakelite First synthetic resin, or plastic.

bathyscaphe Deep-diving underwater vehicle.

bilancetta (hydrostatic balance) Device for measuring the density of an object.

binary number system System of numbers with a base of 2, using the digits 0 and 1; all digital computer code uses the binary number system.

biodiesel Fuel that can be made from different organic materials, including soybeans, palm oil, peanut oil, hemp, and animal and vegetable fats.

biodiversity Measure of the number of living species on earth, of the genetic variety within species, and of the diversity of earth's ecosystems.

biotechnology Manipulation of living organisms to produce something new, including pest-resistant crops, pharmaceuticals, and bacterial strains.

blood transfusion Drawing blood from a healthy person and transferring it to another person.

botany Branch of biology that studies plants.

camera obscura Darkened room with a small hole in one wall that permits the display of images on the opposite wall.

canal lock Gated section of a canal used to change the water level gradually.

casting Forming objects from molten metal.

cataract Condition in which the lens of the eye becomes cloudy, leading to blurry, distorted vision, and, if left untreated, blindness.

cathode-ray tube (CRT) Television-like vacuum tube that displays images when its surface is struck by electron beams.

cellulose Naturally abundant carbohydrate made of long chains of interconnected sugar molecules; constitutes the main ingredient of plant-cell walls.

centrifugal compressor (centrifugal chiller) Device that uses refrigerants to cool and blow air; the basis for the modern air conditioner and many other refrigeration systems.

centrifugal force Apparent force that pushes an object outward when it is spun in a circle.

chlorofluorocarbons (CFCs) Gaseous compounds containing carbon, chlorine, fluorine, and sometimes hydrogen; used in a number of household products and in some manufacturing processes; believed to be a cause of ozone depletion.

chronometer Highly accurate clock (or watch) with which sailors could calculate longitude in the 18th century.

codex Tablet on which religious texts were written; a forerunner of the modern book.

Cold War Post–World War II hostility and competition between the Soviet Union and the United States that ended with the collapse of the Soviet Union in 1991.

computed tomography (CT) scanner Medical device that uses radiation to make detailed images of the organs and other interior structures of a body.

conductor A material, such as metal, capable of transmitting heat, electricity, or sound.

conservation Management of a natural resource or ecosystem in a way that prevents destruction or pollution and protects the surrounding environment.

convergence In reference to modern computing, a gradual coming together of computers and other communication technologies.

cuneiform Earliest known written language; incised on clay tablets invented around 3500 BCE by the Babylonians living in a region now part of Iraq.

daguerreotype Earliest version of the photograph; developed by Louis Daguerre and colleague Joseph-Nicéphore Niépce.

deforestation Clearing of forests and subsequent destruction of the area's ecosystems.

deoxyribonucleic acid (DNA) Large, complex molecule containing genetic information found in the nucleus of a cell; also contains other chemical signals that instruct a cell to manufacture proteins.

diabetes Medical condition in which the body fails to produce sufficient quantities of the hormone insulin.

dialysis machine Device that cleans waste products from the blood, a function normally performed by the kidneys.

differentiation Processes by which the cells of an immature organism take on specialized functions.

digital technology Type of computerized electronics that processes information in the form of numbers.

diorama Depictions of dramatic or famous events drawn or painted on translucent paper that is lit from behind; viewed by paying audiences in the early 1800s.

direct current Irreversible electric current that flows in only one direction.

DNA cloning (recombinant DNA technology) Process in which a single DNA sequence from one organism is isolated and multiplied in living cells.

Doppler radar Type of detection system based on radio waves; used to forecast the weather.

dynamics Science concerned with the behavior and impact of moving bodies.

ecologist Scientist who studies the relationship between organisms and their surrounding environment.

economics Social science that concerns itself with the production, distribution, and consumption of goods and services.

ecosystem System containing all the living and nonliving components of a particular area.

electrical circuit Closed path through which electricity flows.

electric motor Motor that uses electricity to generate the magnetism that makes the central part of the motor rotate so it can move a machine.

electrogasdynamics Study of the behavior of electrically charged gas particles, or ions, when gas is in motion.

electrolyte Chemical in the casing of a battery that conducts electricity.

electromagnetic radiation Flow of energy in the form of the electric and magnetic fields; composed of electromagnetic waves such as radio waves, visible light, and x-rays.

electromagnetic waves Waves created by the oscillations of electric and magnetic fields.

electromagnets Magnets that can be switched on and off by electricity.

electron Elementary particle with a negative charge; present in all atoms.

electrophotography Process used in copying images, later called xerography.

elementary particle Particle smaller than an atom that is not composed of smaller parts.

elocution Art of speaking clearly.

embryo Stage of development of a plant or animal following fertilization of an egg cell.

embryology Branch of biology that studies embryos and embryo development.

emulsion In photography, a substance that, when placed as a thin film onto a material, makes the material sensitive to light.

Escherichia coli (**E. coli**) Type of bacteria that lives in the digestive tract of most mammals, including humans; can be toxic when present in larger numbers than usual.

eugenics Controversial forerunner of modern genetics; it claimed to demonstrate that careful choice of parents could eliminate diseases and improve humankind.

exploratory surgery Type of surgery performed to locate a potential medical problem.

external combustion engine Engine that burns fuel (combusts) outside its cylinders where power is produced.

fermentation Anaerobic breakdown of energy-rich compounds by bacteria.

fiber optics Extremely thin glass fibers used to transmit light by internal reflections, thus creating images or sending information.

flip-flop circuit Circuit that has the ability to go back and forth between two states.

forensic science Branch of science that uses medical knowledge in various kinds of analyses, such as in historical research and in criminal investigations.

fossil fuels Fuels including coal, oil, and gas that are formed from organic matter that has decomposed and often been compressed over millions of years.

gamma rays Radiation emitted by photons (particles of energy that carry light) of higher energy than that of an x-ray, emitted by a radioactive substance.

genetic engineering Introduction of genetic material, typically DNA, from one or more organisms into another organism, causing a change in characteristics of the receiving organism.

geneticist Biologist who studies the heredity and variation of organisms.

genetics Scientific study concerned with heredity and variation of organisms.

geocentric theory (Ptolemaic theory) Theory stating that the earth is at the center of the universe and that the sun revolves around it; originally proposed by Egyptian astronomer and mathematician Ptolemy.

globalization Trends relating to the movement of jobs, goods, services, money, and ideas around the world.

Global System for Mobile Telecommunications (GSM) European cellular phone system that utilizes digital technology; launched in 1980s by a group of telephone manufacturers.

global warming Gradual rise in earth's temperature that is slowly changing the world's climate.

green revolution Movement that began in the 1960s with the introduction of farming methods—including the introduction of new strains of wheat and rice, along with the application of chemical fertilizers and pesticides—that allowed areas of the world with low crop yields to become self-sufficient in food crops.

gyroscope Heavy, spinning wheel used for a number of purposes, including stabilizing rockets during flight.

heat engine Machine that extracts the power locked in fuels such as coal and gasoline by alternately heating and cooling such fuels.

heliocentric theory (Copernican theory) Theory stating that the sun is the center of the universe and that the earth revolves around it; originally proposed by Nicolaus Copernicus.

horticulturalist Scientist who researches the growing of fruits, vegetables, flowers, and other plants.

Human Genome Project Global scientific effort that mapped all the information in human DNA.

hydrocarbons Complex chemicals formed from the elements hydrogen and carbon.

hydroelectric power plant Power plant that produces electricity using energy created by flowing water.

hydrogen bomb (fusion bomb) Powerful bomb in which isotopes of hydrogen are fused together, releasing great amounts of energy.

hydrostatics Field of science that studies fluids at rest, including buoyancy—the ability of materials less dense than water to float.

hypertext Way of connecting documents on a computer system with highlighted words or phrases; integral to the functionality of the World Wide Web.

HyperText Markup Language (HTML) Structured format used to create documents on the World Wide Web.

HyperText Transfer Protocol (HTTP) Uniform system used to send and receive information over the Internet.

induction telegraphy System invented in the 1880s that uses electromagnetism to communicate with moving elevator or train cars.

Industrial Revolution Large and rapid changes in an economy and a society—as in England in the late 18th century and in the United States in the mid-19th century—caused by the introduction of improved, power-driven machinery and mass production.

information technology Development and use of computers for processing data.

infusion pump Device that administers medications directly into a patient's circulatory system.

insoluble Incapable of being dissolved.

insulator Any material, such as rubber, that is a poor conductor of heat, electricity, or sound.

integrated circuit Miniature electronic circuit on the surface of a semiconducting material, such as silicon.

internal combustion engine Engine that burns fuel (combusts) inside its cylinders, where power is produced.

in vitro fertilization (IVF) Medical procedure in which eggs are taken from a female and fertilized outside the body; the resulting embryo is then placed in the uterus of the same or a different female with the hope of implantation and subsequent pregnancy.

ion Positively or negatively charged atom resulting from the gain or loss of one or more electrons.

ionosphere Part of the atmosphere that bends radio waves around the earth.

Java Powerful programming language written for the World Wide Web in the 1990s.

kinetic energy Energy associated with an object's motion.

kinetograph A primitive movie camera that could record a series of still photographs onto a length of celluloid film.

kinetoscope Early film projector that played celluloid film of the kinetograph.

laparoscopy Surgical technique using a thin fiber-optic tube (laparoscope) inserted into the abdomen through small incisions to provide light for surgical instruments.

laser Powerful generators of pure light used in a variety of applications, including surgery and land surveying.

latitude North–south position on the globe, marked by parallels, which are imaginary lines placed at regular distances from the equator.

Linux Free computer operating system created by Linus Torvalds and developed by thousands of volunteer programmers.

load-bearing capacity The amount of weight various parts of a building can support.

logarithm Way of performing calculations with mathematical tables before calculators were invented; also, exponent of a number to a given base.

longitude East–west position on the globe, marked by meridians, which are imaginary lines running at regular distances around the globe in a north–south direction and meeting at the north and south poles.

magnetic resonance imaging (MRI) Type of diagnostic test using a magnetic field that builds two- and three-dimensional pictures of the soft tissues in the body to help locate medical problems such as tumors.

magnetron Equipment that generates microwaves.

Manhattan Project U.S. government program (1942–1945) credited with the development of the atomic bomb.

maser Acronym for microwave amplification by stimulated emission of radiation; first device to emit microwaves that were of the same wavelength.

mass production Method of industrial production involving the specialized use of labor, standardized parts, and machinery, which results in large quantities of goods made with the least effort and in the least time.

mechanics Scientific study concerned with energy and forces and their effect on objects.

Mercator projection Type of map that preserves the shape of geographic features and their location relative to one another.

mercerize Process of chemically treating cotton to prevent it from shrinking and to improve its strength and luster.

metallurgy Scientific and technological study of metals.

methane Flammable gas made of carbon and hydrogen.

microbiology Branch of biology that studies microscopic life-forms.

microprocessor (microchip) Tiny silicon chip that contains all the essential components of a computer; it made possible such devices as pocket calculators and digital watches; another name for an integrated circuit.

microwaves Radio waves of very short wavelength.

microwave spectroscopy Use of microwaves to study molecules.

molecule Two or more atoms joined by chemical bonds and behaving as one individual unit in chemical reactions.

monoculture Cultivation of a single crop across a wide area of land; usually associated with planting the same crop in successive growing seasons.

Morse code Pattern of short (dots) and long (dashes) electrical pulses that indicate different letters of the alphabet; used for telegraphing messages.

movable type Printing method in which many tiny blocks, each capable of printing one letter, were moved around in a frame or form; these separate blocks could be used repeatedly to print different words; its invention revolutionized printing and education, and allowed the preservation and spread of culture and learning.

nanotechnology System of developing new materials by manipulation of individual atoms and molecules.

nephrology Scientific and medical study of kidneys.

neuroscience Scientific study concerned with the brain and nervous system.

neutron Subatomic particle with no electric charge, located in an atom's nucleus.

nitroglycerin Highly explosive liquid chemical used in dynamite.

nuclear fission Splitting of the nuclei of two or more atoms, which results in a release of energy.

nuclear fusion Joining of the nuclei of two or more atoms, which results in the conversion of some of their mass into energy.

nuclear magnetic resonance (NMR) technology Absorption of very high-frequency radio waves by certain atomic nuclei that are subjected to an appropriately strong stationary magnetic field.

nuclear transfer Process of cloning living entities in which the nucleus of one cell is replaced by the nucleus of another cell; often used to clone living organisms.

nucleus Positively charged core of an atom, containing neutrons and protons.

nylon Highly durable synthetic material made into a variety of forms, such as fiber, bristles, or sheets; has many different applications, as in hosiery and thread to stitch wounds together.

oceanography Scientific study of oceans and their waters, marine life, and resources.

open source Uncopyrighted, original computer software, or source code, freely available to the public.

ophthalmology Field of medical science dealing with the study and treatment of the structure, function, and diseases of the eye.

optics Scientific study concerned with the behavior of light.

organic chemistry Branch of chemistry that studies carbon and its compounds.

parabola Inverted U-shaped path.

pasteurization Invented by Louis Pasteur; method of preserving food, often a dairy product, by destroying harmful organisms.

patent Certificate that confers exclusive rights to an invention or design for a set period.

penicillin An antibiotic used to treat many diseases caused by microorganisms.

pesticides Chemical agents used to eliminate insects or pests.

petrograms Earliest preserved images of drawings made on cave walls.

philanthropy Contribution, typically financial, to charitable causes.

photoconductivity Increased electrical conductivity of substances created when light is passed through them.

photoelectric theory (photoelectric effect) Albert Einstein's theory describing the connection between electricity and light, explaining how light can be used to generate electricity.

photon Particle of energy by which light is carried.

photosynthesis Process by which green plants convert the sun's energy into energy-rich substances.

photovoltaic panel Modern device that converts the sun's energy into electrical power.

physics Mathematic and scientific study of matter and energy and how the two interact.

physiology Scientific study concerned with the activities and functions of major internal structures of the body.

pi (π) A value expressing the ratio of the circumference of a circle to its diameter; first approximated by Archimedes of Syracuse.

planar process Invention that allowed hundreds, thousands, and eventually millions of transistors to be placed on a single silicon chip.

planetarium Model of the sun and planets.

plant pathology Study of disease in plants.

plasma Pale, watery part of the blood wherein the red blood cells, which carry oxygen and other vital supplies to different parts of the body, are suspended.

plasmids Loops of DNA found in one-celled organisms that are not contained in a cell's nucleus.

polychloroprene (neoprene) First synthetic rubber, perhaps most widely known as the material used to make wet suits.

polymath Person knowledgeable in many different fields.

polymerization Building large molecules from many smaller ones based on the chemical element carbon.

polymers Huge carbon-based molecules made from thousands of separate atoms.

polytechnic School devoted to applied sciences and technical skills.

proton Positively charged particle in the nucleus of an atom.

prototype Test version of an invention.

radar Type of navigational device used by ships and airplanes; it emits powerful radio waves to detect the location of objects such as airplanes and storms and processes their reflections for display; acronym for **ra**dio **d**efense **a**nd **r**anging.

radiation Any energy emitted in the form of particles or waves.

radio Waves created by oscillations of electric and magnetic fields.

radioactivity High-energy rays or tiny particles given off by unstable atoms.

radium Chemical element discovered in 1898 by Marie Curie; used to develop first radiation treatments for cancer.

rayon Soft cloth made of dissolved cellulose.

red blood cells Blood cells that carry oxygen and other life-sustaining substances to different parts of the body.

reduction Chemical process that restores electrons to atoms.

Reformation Period of religious upheaval in the sixteenth century when groups broke from Catholicism and established Protestant churches.

Renaissance Period of history roughly between 1400 and 1600 when many advances in art, religion, and science were made.

scientific revolution Period beginning in the late 1500s, when science began to replace religion as the most credible explanation for phenomena of the everyday world.

self-contained underwater breathing apparatus (scuba) Underwater breathing equipment that delivers an air supply from a tank to an underwater diver's mouthpiece; also known as the Aqua-Lung.

semiconductors Materials such as silicon and germanium that allow electricity to flow only when impurities are added to them; widely used in electronic devices.

shellac Natural compound extracted from the Asian lac beetle; used for coating and sealing wood.

Silicon Valley Industrial region located in California's San Francisco Bay area that has become a center for electronics and computer companies since the mid-20th century.

solar cells Devices that use semiconductors such as silicon to turn sunlight into electricity.

solar energy Energy derived from the sun's radiation, used to create heat, electricity, or a chemical reaction.

sonar Navigational device using sound waves to detect objects underwater; acronym for **so**und **na**vigation **r**anging.

spread spectrum technology Radio-based guidance systems that switch from frequency to frequency; originally utilized by the military, then adopted for civilian use in wireless networks and cellular phones.

static electricity Electrical charge that accumulates in one place—such as under thunderclouds—and is produced through friction or induction.

statistics Field involving the collection and analysis of data; often applied to the study of human behavior.

steam engine Type of heat engine that burns coal to boil water, then uses the resulting steam to drive railroad locomotives and other machines.

steam turbine Engine that turns steam into electricity.

stem cell Undifferentiated cell with the ability to produce other cells that take on specialized functions.

stimulated emission Process in which excited molecules are bombarded with additional energy, causing the molecules to release more energy than they had absorbed.

stock ticker Early printer that recorded stock prices arriving as telegraph messages.

subatomic particles Tiny components, including protons, neutrons, and electrons, that make up atoms.

superpolyamides Group of chemicals, each with a slightly different arrangement of atoms that gives it different properties.

supersonic Faster than sound.

synthetic material Chemically produced material not occurring in nature.

taxidermy Process of preparing, stuffing, and mounting dead animals for display, especially vertebrates.

telecommunications Means of communication over long distances; includes telephone, radio, and television.

telegraph Communication device that used electric cables and Morse code to send messages at high speed, one letter at a time.

textiles Cloth made from natural or artificial fibers.

theory of relativity Theory proposed by Albert Einstein stating that matter and energy are the same and that even a tiny amount of matter can produce an enormous amount of energy.

thermodynamics Scientific study of how heat moves.

thermoscope Early version of a thermometer-like instrument developed by Galileo Galilei.

thermosetting The making of permanently firm plastic or other material that cannot melt or be reshaped when heated.

totipotency Condition in which a cell, such as a zygote, has the potential and ability to become any kind of specialized cell.

transistors Microscopic electronic switches that are the electronic "brains" (microprocessors or microchips) of digital watches, personal computers, and other electronic devices.

Uniform Resource Locator (URL) Unique name address given to each document on the Internet; makes document easily accessible.

vaccination Treatment that injects individuals with a weakened form of a virus to produce immunity (the ability of the body to halt the development of the full-blown disease).

vacuum tube Small electronic component that amplifies weak electronic currents; credited with making radio a worldwide success.

vulcanization Process that hardens rubber by heating it and adding sulfur.

waterborne disease Any disease contracted by drinking or coming into contact with unsanitary water, including cholera; one of the leading causes of death in many parts of the developing world.

waterwheel Large open-paddle wheel usually made from wood that is turned by the force of a river, harnessing the water's power for use in a mill or other machine.

World Wide Web An Internet system that allows easier navigation of the network through graphical user interfaces and inter-linked hypertext documents.

x-ray Form of radiation used in medicine to get images of internal body structures.

zoology Scientific study of animal life, including the classification, properties, and behaviors of animals.

zygote A fertilized cell.

RESOURCES FOR FURTHER STUDY

GENERAL RESOURCES ON INVENTORS AND INVENTIONS

Books

Altman, Linda Jacobs. *Women Inventors*. New York: Facts On File, 1997.

Amram, Fred. *African-American Inventors*. Mankato, MN: Capstone, 1996.

Bender, Lionel. *Eyewitness: Invention*. New York: Dorling Kindersley, 2005.

Bridgman, Roger. *1,000 Inventions and Discoveries*. New York: Dorling Kindersley, 2006.

Brodie, James Michael. *Created Equal: The Lives and Ideas of Black American Innovators*. New York: Morrow, Bill Adler Books, 1993.

Brown, David E. *Inventing Modern America: From the Microwave to the Mouse*. Cambridge, MA: MIT Press, 2001.

Caney, Stephen. *Stephen Caney's Invention Book*. New York: Workman, 1985.

Carwell, Hattie. *Blacks in Science: Astrophysicist to Zoologist*. Hicksville, NY: Exposition, 1977.

Casey, Susan. *Women Invent: Two Centuries of Discoveries That Have Helped Our World*. Chicago: Chicago Review Press, 1997.

Fouché, Rayvon. *Black Inventors in the Age of Segregation: Granville T. Woods, Lewis H. Latimer, and Shelby J. Davidson*. Baltimore, MD: Johns Hopkins University Press, 2003.

Fucini, Joseph J., and Suzy Fucini. *Entrepreneurs*. Boston: G.K. Hall, 1985.

Henderson, Susan K. *African-American Inventors II*. Mankato, MN: Capstone, 1998.

Jones, Charlotte. *Mistakes That Worked*. New York: Doubleday, 1994.

Jones, Charlotte Foltz. *Accidents May Happen*. New York: Delacorte, 1996.

Kirkham, Pat, ed. *Women Designers in the USA: 1900–2000*. New Haven, CT: Yale University Press, 2000.

Klein, Aaron E. *The Hidden Contributors: Black Scientists and Inventors in America*. New York: Doubleday, 1971.

Levy, Richard. *The Complete Idiot's Guide to Cashing In on Your Inventions*. Indianapolis, IN: Alpha, 2001.

Macdonald, Anne L. *Feminine Ingenuity: Women and Invention in America*. New York: Ballantine, 1992.

Smith, Roger, ed. *Inventions and Inventors*. Pasadena, CA: Salem, 2002.

Sobey, Ed. *How to Enter and Win an Invention Contest*. Hillside, NJ: Enslow, 1999.

Stanley, Autumn. *Mothers and Daughters of Invention*. Piscataway, NJ: Rutgers University Press, 1995.

Sullivan, Otha Richard. *Black Stars: African American Women Scientists and Inventors*. New York: Wiley, 2001.

Thimmesh, Melissa Sweet. *Girls Think of Everything: Stories of Ingenious Inventions by Women*. Boston: Houghton Mifflin, 2000.

Tucker, Tom, and Richard Loehle. *Brainstorm!: The Stories of Twenty American Kid Inventors*. New York: Farrar, Straus, and Giroux, 1998.

Van Sertima, Ivan, ed. *Blacks in Science: Ancient and Modern.* New Brunswick, NJ: Transaction, 1983.

Vare, Ethlie Ann, and Greg Ptacek. *Mothers of Invention.* New York: Morrow, 1988.

———. *Patently Female: From AZT to TV Dinners, Stories of Women and Their Breakthrough Ideas.* New York: Wiley, 2002.

Von Oeck, Roger. *A Whack on the Side of the Head.* New York: Warner Business, 1998.

Williams, Trevor. *A History of Invention.* New York: Time Warner, 2004.

Wulffson, Don. *The Kid Who Invented the Popsicle, and Other Surprising Stories about Inventions.* New York: Puffin, 1999.

Web sites

American Inventor
>   Web site of the ABC Television inventing contest.
>   http://abc.go.com/primetime/americaninventor

Black Inventor Online Museum
>   A collection of biographies hosted by Adscape International.
>   http://www.blackinventor.com/pages/lloydhall.html

DuPont Heritage
>   A historical exhibition from the chemical company.
>   http://heritage.dupont.com

European Patent Office
>   Responsible for granting patents in the European Union.
>   http://www.european-patent-office.org

Explore Invention at the Lemelson Center
>   Explores the history of invention and encourages young people to create new inventions.
>   http://invention.smithsonian.org/home

Google: Patents
>   Search the entire collection of U.S. patents from the 1790s to 2006.
>   http://www.google.com/patents

Great Sites for Young Inventors
>   How to get involved in inventing.
>   http://inventors.about.com/od/kidactivities

IEEE Virtual Museum
>   Information about and activities for exploring electricity, electronics, and modern communications technology.
>   http://www.ieee-virtual-museum.org

Innovative Lives
>   Stories of invention and innovation from the Smithsonian Institution.
>   http://www.invention.smithsonian.org/centerpieces/ilives

Invent Help
>   Practical advice for inventors, including lists of inventing contests.
>   http://www.inventhelp.com

Invention at Play
> Discover how play helps creativity and invention.
> http://www.inventionatplay.org

Inventors from About.com
> A Web site that describes many familiar inventions.
> http://inventors.about.com

Kid Inventor Challenge
> An annual contest run by Wild Planet Toys.
> http://www.kidinventorchallenge.com

Lemelson-MIT Prize
> Details of the "Oscar for inventors," with lists of past winners.
> http://web.mit.edu/INVENT/a-prize.html

National Inventors Hall of Fame
> Profiles of many well-known inventors.
> http://www.invent.org

Rube Goldberg
> Includes examples of Rube Goldberg's cartoons and details of the Rube Goldberg Machine Contest.
> http://www.rube-goldberg.com

U.S. Patent and Trademark Office
> Home page of the USPTO includes a patent search page and a children's guide to patents.
> http://www.uspto.gov

Young Inventors International
> A nonprofit group dedicated to helping budding inventors.
> http://www.younginventorsinternational.com

BUILDINGS AND MATERIALS

Books

Cadbury, Deborah. *Dreams of Iron and Steel*. New York: Fourth Estate, 2004.

Levy, Matthys, and Mario Salvadori. *Why Buildings Fall Down*. New York: Norton, 1994.

Macaulay, David. *Building Big*. New York: Houghton Mifflin, 2004.

Misa, Thomas J. *A Nation of Steel: The Making of Modern America, 1865–1925*. Baltimore, MD: Johns Hopkins University Press, 1995.

Newby, Frank, ed. *Early Reinforced Concrete*. Aldershot, UK: Ashgate, 2001.

Sabbagh, Karl. *Skyscraper: The Making of a Building*. New York and London: Penguin, 1989.

Wilkinson, Philip. *Eyewitness: Building*. New York: Dorling Kindersley, 2000.

Web sites

Architects and Buildings: Solar Houses
>Information on the solar architecture project at MIT.
>http://libraries.mit.edu/guides/subjects/architecture/architects/solar/index.html

Building Big
>Companion to the PBS series on buildings.
>http://www.pbs.org/wgbh/buildingbig

Great Buildings Online
>Information and 3-D models of the world's most famous buildings.
>http://www.greatbuildings.com

The History of Concrete
>A time line by the University of Illinois Urbana-Champaign.
>http://matse1.mse.uiuc.edu/concrete/hist.html

History of Reinforced Concrete
>Information from the International Association of Bridge, Structural, Ornamental, and Reinforcing Iron Workers.
>http://www.ironworkers.org/ht/display/ArticleDetails/i/1401

How a Blast Furnace Makes Iron
>An animated game from BBC Education.
>http://www.bbc.co.uk/history/games/blast/blast.shtml

How Iron and Steel Work
>An introduction to the basics of ironmaking and steelmaking.
>http://science.howstuffworks.com/iron.htm

PBS.org: Building Big: All about Skyscrapers
>Educational site about skyscrapers and their history.
>http://www.pbs.org/wgbh/buildingbig/skyscraper

Polymers—They're Everywhere
>A National Geographic educational site about plastics and polymers.
>http://www.nationalgeographic.com/resources/ngo/education/plastics

Skyscraper Museum
>Online home of New York City's Skyscraper Museum.
>http://www.skyscraper.org/home_flash.htm

The Story of Rubber
>An online exhibition from the Polymer Science Learning Center.
>http://www.pslc.ws/macrog/exp/rubber/menu.htm

Super Bridge: Suspension Bridge
>An exhibit on suspension bridges by PBS's *NOVA*.
>http://www.pbs.org/wgbh/nova/bridge/meetsusp.html

CLOTH AND APPAREL

Books

Carlson, Laurie. *Queen of Inventions: How the Sewing Machine Changed the World.* Brookfield, CT: Millbrook, 2003.

Gilpin, Daniel. *Food and Clothing (The History of Invention)*. New York: Facts On File, 2004.

Hamilton, Sue L. *Clothing: A Pictorial History of the Past One Thousand Years*. Edina, MN: ABDO, 2000.

Harris, Jennifer. *5,000 Years of Textiles*. London: British Museum, 2004.

Rowland-Warne, L. *Eyewitness: Costume*. New York: Dorling Kindersley, 2000.

Schoesser, Mary. *World Textiles: A Concise History*. New York: Thames and Hudson, 2003.

Steele, Philip. *Clothes and Crafts in Victorian Times*. Milwaukee, WI: Gareth Stevens, 2000.

Web sites

Encyclopedia of British History: Textile Industry
Articles about important inventors and inventions from the textile revolution.
http://www.spartacus.schoolnet.co.uk/Textiles.htm

Spinning the Web: The Story of the Cotton Industry
A collection of digitized items from libraries and archives in northwest England, developed by the Manchester Library and Information Service.
http://www.spinningtheweb.org.uk

COMMUNICATIONS

Books

Bridgman, Roger. *Eyewitness: Technology*. New York: Dorling Kindersley, 1998.

Briggs, Asa, and Peter Burke. *A Social History of the Media: From Gutenberg to the Internet*. 2nd ed. Malden, MA: Polity, 2005.

Crowley, David, and Paul Heyer. *Communication in History: Technology, Culture, and Society*. Boston: Allyn and Bacon, 2002.

Davidson, Cathy N. *Reading in America: Literature and Social History*. Baltimore, MD: Johns Hopkins University Press, 1989.

Fischer, Claude S. *America Calling: A Social History of the Telephone to 1940*. Berkeley: University of California Press, 1992.

Goldsmith, Mike. *Fantastic Future*. New York: Scholastic, 2004.

Lundmark, Tobjorn. *Quirky Qwerty: A Biography of the Typewriter and Its Many Characters*. London: Penguin, 2003.

Marvin, Caroline. *When Old Technologies Were New: Thinking about Electric Communication in the Late Nineteenth Century*. New York: Oxford University Press, 1990.

McKitterick, David. *Print, Manuscript, and the Search for Order, 1450–1830*. New York: Cambridge University Press, 2003.

Newcomb, Horace, ed. *Museum of Broadcast Communications, Encyclopedia of Television*. Chicago: Fitzroy Dearborn, 1997.

Standage, Tom. *The Victorian Internet*. New York: Berkley, 1999.

Underwood, Doug. *From Yahweh to Yahoo: The Religious Roots of the Secular Press*. Champaign: University of Illinois Press, 2002.

Wallace, Aurora. *Newspapers and the Making of Modern America: A History.* Westport, CT: Greenwood, 2005.

Woodford, Chris. *Communication and Computers.* New York: Facts On File, 2004.

Wright, Russell O. *Chronology of Communication in the United States.* Jefferson, NC: McFarland, 2004.

Web sites

History of Communication Research Bibliography
    A searchable database of thousands of citations on communications history.
    http://www.historyofcommunicationresearch.org

International Printing Museum
    Information about printing methods and printing presses through history.
    http://www.printmuseum.org

U.S. Marconi Museum
    A history of radio in the United States.
    http://www.marconiusa.org

Virtual Antique Typewriter Museum
    A large collection of historical photographs and documents.
    http://www.typewritermuseum.org

COMPUTERS

Books

Campbell-Kelly, Martin, and William Aspray. *Computer: A History of the Information Machine.* New York: Basic Books, 1996.

Cringely, Robert X. *Accidental Empires: How the Boys of Silicon Valley Make Their Millions, Battle Foreign Competition, and Still Can't Get a Date.* New York: HarperBusiness, 1996.

Gehani, Narain. *Bell Labs: Life in the Crown Jewel.* Summit, NJ: Silicon, 2003.

Gralla, Preston. *Online Activities for Kids.* New York: Wiley, 2001.

Hiltzik, Michael. *Dealers of Lightning: Xerox PARC and the Dawn of the Computer Age.* New York: Collins, 2000.

Levy, Steven. *Hackers: Heroes of the Computer Revolution.* New York: Penguin, 2001.

Mackinnon, Mairi. *Usborne Guide to the Internet.* New York: Usborne, 2002.

McAfee, John. *Computer Viruses, Worms, Data Diddlers, Killer Programs, and Other Threats to Your System.* New York: St. Martin's, 1989.

Moody, Glynn. *Rebel Code: Linux and the Open Source Revolution.* New York: Penguin, 2002.

Shurkin, Joel. *Engineering the Mind: A History of the Computer.* New York: Norton, 1984.

Smith, George C. *The Virus Creation Labs: A Journey into the Underground.* Tucson, AZ: American Eagle, 1994.

Williams, Brian. *Computers: Great Inventions.* Portsmouth, NH: Heinemann, 2001.

Wingate, Philippa. *The Usborne Book of the Internet*. London: Usborne, 2002.

Woodford, Chris. *History of Invention: Communications and Computers*. New York: Facts On File, 2004.

Wurster, Christian. *Computers: An Illustrated History*. New York: Taschen, 2002.

Web sites

Computer History Museum
A California museum whose Web site includes online exhibits, photographs, and a time line.
http://www.computerhistory.org

IEEE Virtual Museum
The history of electricity, electronics, and computers.
http://www.ieee-virtual-museum.org

Internet Storm Center
A site that continually assesses threats to the Internet from viruses and other attacks.
http://isc.incidents.org

The Journey Inside
An interactive Web site from Intel about integrated circuit technology.
http://www.intel.com/education/journey/index.htm

Landmarks in Digital Computing
A history of computing from the National Museum of American History.
http://www.nasm.si.edu/research/dsh/LDC/publicit.html

Open Directory Project: Kids and Teens: Computers and Internet
A broad collection of resources for young people.
http://dmoz.org/Kids_and_Teens/Computers/Internet

OpenSources: Voices from the Open Source Revolution
Interviews with open-source pioneers, including Linus Torvalds.
http://www.oreilly.com/catalog/opensources/book/toc.html

Symantec Security Response Center
An up-to-date assessment of latest viruses.
http://securityresponse.symantec.com

Triumph of the Nerds
A PBS Web site about the development of personal computers in California in the 1980s.
http://www.pbs.org/nerds

ENERGY AND POWER

Books

Behrman, Daniel. *Solar Energy: The Awakening Science*. Boston: Little, Brown, 1976.

Butti, Ken, and John Perlin. *A Golden Thread: 2500 Years of Solar Architecture and Technology*. Palo Alto, CA: Cheshire, 1980.

Challoner, Jack. *Eyewitness: Energy*. New York: Dorling Kindersley, 2000.

Deford, Debra. *The Steam Engine*. New York: World Almanac, 2005.

Ramage, Janet. *Energy: A Guidebook*. Oxford and New York: Oxford University Press, 1997.

Sutcliffe, Andrea. *Steam: The Untold Story of America's First Great Invention*. New York: Palgrave Macmillan, 2004.

Woodford, Chris. *Energy and Electricity*. New York: Facts On File, 2004.

———. *Power and Energy*. New York: Facts On File, 2004.

———. *Routes of Science: Electricity*. Farmington Hills, MI: Blackbirch, 2004.

———, and Martin Clowes. *Atoms and Molecules: Routes of Science*. Farmington Hills, MI: Blackbirch, 2004.

Web sites

Center for Alternative Technology
A comprehensive introduction to renewable energy.
http://www.cat.org.uk

Energy Information Administration: Kid's Page
Information, games, and activities from the U.S. Department of Energy.
http://www.eia.doe.gov/kids

Energy Quest: Energy Education
An introduction from the California Energy Commission.
http://www.energyquest.ca.gov

How Hydropower Plants Work
A look at hydroelectric technology from the Howstuffworks Web site.
http://www.howstuffworks.com/hydropower-plant.htm

Hydrogen Hopes
A presentation on hydrogen technology, including Stanford Ovshinsky's work, from PBS.
http://www.pbs.org/saf/1506/index.html

Kids' Zone: Nuclear Energy
Atomic Energy of Canada's interactive guide to the energy locked inside atoms.
http://www.aecl.ca/kidszone/atomicenergy/nuclear/puzzle.asp

Socket to Me! How Electricity Came to Be
The story of electricity and the familiar household appliances it spawned.
http://www.ieee-virtual-museum.org

Steam Engine Library
A collection of historical documents relating to the history of the steam engine.
http://www.history.rochester.edu/steam

Students' Corner: Nuclear Energy
A guide by the U.S. Nuclear Regulatory Commission.
http://www.nrc.gov/reading-rm/basic-ref/students.html

U.S. Bureau of Reclamation: Dams
Information about the dams and hydroelectric plants in the United States.
http://www.usbr.gov/dataweb/dams/index.html

World Commission on Dams
    Detailed information about the economic benefits and environmental drawbacks
    of hydroelectric dams.
    http://www.dams.org

ENTERTAINMENT

Books

Abramson, Albert. *The History of Television, 1942 to 2000*. Jefferson, NC: McFarland,
    2003.
Ardley, Neil. *Eyewitness: Music*. New York: Dorling Kindersley, 2004.
Buckingham, Alan. *Eyewitness: Photography*. New York: Dorling Kindersley, 2004.
Cartmell, Robert. *The Incredible Scream Machine: A History of the Roller Coaster*.
    Fairview Park, OH: Amusement Park, 1987.
Coleman, Mark. *Playback: From the Victrola to MP3, 100 Years of Music, Machines, and
    Money*. Cambridge, MA.: Da Capo, 2003.
Editors of *LIFE*. *LIFE: 100 Photographs That Changed the World*. New York: LIFE, 2003.
Florida, Richard. *The Rise of the Creative Class: And How It's Transforming Work,
    Leisure, Community, and Everyday Life*. New York: Basic, 2004.
Johnson, Neil. *Photography Guide for Kids*. Washington, DC: National Geographic,
    2001.
Kent, Stephen L. *The Ultimate History of Video Games: From Pong to Pokemon*. New
    York: Three Rivers, 2001.
Newcomb, Horace, ed. *Museum of Broadcast Communications, Encyclopedia of
    Television*. Chicago: Fitzroy Dearborn, 1997.
O'Brien, Lisa. *Lights, Camera, Action!: Making Movies and TV from the Inside Out*.
    Ontario, Canada: Maple Tree, 1998.
Parkinson, David. *The Young Oxford Book of the Movies*. New York: Oxford, 1997.
Poole, Steven. *Trigger Happy: Videogames and the Entertainment Revolution*. New York:
    Arcade, 2004.
Rosenblum, Naomi. *A World History of Photography*. New York: Abbeville, 1997.
Stashower, Donald. *The Boy Genius and the Mogul: The Untold Story of Television*. New
    York: Broadway, 2002.
Wiese, Jim. *Movie Science: 40 Mind-Expanding, Reality-Bending, Starstruck Activities for
    Kids*. New York: Wiley, 2001.

Web sites

A History of Vinyl
    A BBC exhibit on the development of the phonograph record.
    http://www.bbc.co.uk/music/features/vinyl
American Museum of Photography
    Extensive information about the history of photography.
    http://www.photography-museum.com

Color TV's 50th Anniversary

NPR stories and videos marking half a century of color television.

http://www.npr.org/templates/story/story.php?storyId=1789944

Early Cinema

A look at the pioneers of movies and moviemaking.

http://www.earlycinema.com

Early Television Museum

Essays about and photographs of early television.

http://www.earlytelevision.org/index.html

Inventing Entertainment

A collection of Thomas Edison's sound recordings and movies at the Library of Congress.

http://memory.loc.gov/ammem/edhtml/edhome.html

The Race to Video

A history of the development of video recording from *Invention & Technology* magazine.

http://www.americanheritage.com/articles/magazine/it/1994/2/1994_2_52.shtml

U.S. Marconi Museum

A history of radio in the United States.

http://www.marconiusa.org

FOOD AND AGRICULTURE

Books

Borlaug, Norman E. *Land Use, Food, Energy, and Recreation*. Aspen, CO.: Aspen Institute for Humanistic Studies, 1983.

Francis, Frederick, ed. *Wiley Encyclopedia of Food Technology*. New York: Wiley, 1999.

Gilpin, Daniel. *Food and Clothing (The History of Invention)*. New York: Facts On File, 2004.

Goldberg, Jake. *The Disappearing American Farm*. New York: Franklin Watts, 1996.

Halley, Ned. *Eyewitness: Farm*. New York: Dorling Kindersley, 2000.

Jango-Cohen, Judith. *The History of Food (Major Inventions through History)*. Minneapolis, MN: Twenty-First Century, 2004.

Miller, Char, ed. *Atlas of U.S. and Canadian Environmental History*. New York: Routledge, 2004.

Thomas, Eric, and Angela Wilkes. *A Farm through Time*. New York: Dorling Kindersley, 2001.

Web sites

Agriculture in the Classroom

Activities, projects, and information from the U.S. Department of Agriculture.

http://www.agclassroom.org

Harvest of Fear

An introduction to the modern biotechnology debate from PBS.

http://www.pbs.org/wgbh/harvest

National Museum of American History: Agriculture
    Items of agricultural history from the Smithsonian Institution's collection.
    http://americanhistory.si.edu/collections
Pew Initiative on Food and Biotechnology
    Information on DNA cloning, its impact on agriculture, and the resulting debates.
    http://pewagbiotech.org

HEALTH AND MEDICINE

Books

Agar, Jon. *Constant Touch: A Global History of the Mobile Phone.* New York: Totem, 2005.

Bridgman, Roger. *Eyewitness: Electronics.* New York: Dorling Kindersley, 2000.

Cule, John. *Timetables of Medicine.* New York: Black Dog and Leventhal, 2000.

Drugs and Society. New York: Marshall Cavendish, 2006.

Levinson, Paul. *Cellphone: The Story of the World's Most Mobile Medium and How It Has Transformed Everything!* New York: Macmillan, 2004.

Link, Kurt. *The Vaccine Controversy: The History, Use, and Safety of Vaccinations.* Westport, CT: Greenwood, 2005.

Parker, Steve. *Eyewitness: Medicine.* New York: Dorling Kindersley, 2001.

Porter, Roy. *The Cambridge Illustrated History of Medicine.* New York: Cambridge University Press, 2001.

————. *Blood and Guts: A Short History of Medicine.* New York: Penguin, 2003.

Tiner, John Hudson. *Exploring the History of Medicine.* Green Forest, AR: New Leaf/ Master, 1999.

Wilmut, Ian, and Roger Highfield. *After Dolly: The Uses and Misuses of Human Cloning.* New York: Norton, 2006.

Wolbarst, Anthony B. *Looking Within: How X-Ray, CT, MRI, Ultrasound, and Other Medical Images Are Created, and How They Help Physicians Save Lives.* Berkeley: University of California Press, 1999.

Web sites

American Society for Microbiology: Stalking the Mysterious Microbe
    Information and activities.
    http://www.microbe.org
Bioethics.net: Cloning
    News and discussion about cloning from the Center for Health Care Ethics.
    http://www.bioethics.net/topics.php?catId=4
Body and Mind
    An educational site about medicine and health.
    http://www.bam.gov

Contemporary Anesthesia and Anesthetics

A scientific explanation of modern general anesthesia and what is known about its molecular mechanisms.

http://www.general-anaesthesia.com/painless.html

Genes Can Be Moved between Species

An exhibit on DNA cloning by the Josiah Macy, Jr., Foundation.

http://www.dnaftb.org/dnaftb/34/concept/index.html

Human Genome Project

Information about the history, science, and benefits of the human genome project.

http://www.ornl.gov/sci/techresources/Human_Genome/home.shtml

KidsHealth

Information, activities, and advice on health issues.

http://www.teenshealth.org/teen

Pasteur Institute

A history of the French laboratory that specializes in research on diseases.

http://nobelprize.org/medicine/articles/jacob

United States National Library of Medicine

Online exhibits about the history of medicine.

http://www.nlm.nih.gov/onlineexhibitions.html

Utopian Surgery: Early Arguments against Anesthesia in Surgery, Dentistry, and Childbirth

A very detailed history of the science of anesthesia, with links to primary historical sources.

http://www.general-anaesthesia.com

HOUSEHOLD INVENTIONS

Books

Carlson, Laurie. *Queen of Inventions: How the Sewing Machine Changed the World.* Brookfield, CT: Millbrook, 2003.

Cowan, Ruth Schwartz. *More Work for Mother: The Ironies of Household Technology from the Open Hearth to the Microwave.* New York: Basic Books, 1985.

Ford, Barbara. *Keeping Things Cool: The Story of Refrigeration and Air Conditioning.* New York: Walker, 1986.

Ierley, Merritt. *Comforts of Home: The American House and the Evolution of Modern Convenience.* Mendham, NJ: Astragal, 1999.

Landau, Elaine. *The History of Everyday Life.* Breckenridge, CO: Twenty-First Century, 2005.

Sparke, Penny, ed. *The Plastics Age: From Bakelite to Beanbags and Beyond.* Woodstock, NY: Overlook, 1993.

Wolke, Robert L. *What Einstein Told His Cook: Kitchen Science Explained.* New York: Norton, 2002.

Web sites

History of Home Inventions
    A collection of links to articles about a wide variety of household inventions.
    http://inventors.about.com/library/inventors/blhome.htm
Socket to Me! How Electricity Came to Be
    The story of electricity and the familiar household appliances it spawned.
    http://www.ieee-virtual-museum.org

MILITARY AND WEAPONRY

Books

Brown, G. I. *The Big Bang: A History of Explosives*. Glostershire, UK: Sutton, 2005.
Chorlton, Windsor. *Weapons and Technology of World War II*. New York: Heinemann, 2002.
Jones, Glyn. *The Jet Pioneers: The Birth of Jet-Powered Flight*. London: Methuen, 1989.
Parsons, Jayne. *Eyewitness: World War I*. New York: Dorling Kindersley, 2001.
Richie, John. *Designing the Tools of War*. Minneapolis, MN: Oliver, 2000.
Sheehan, Sean. *The Technology of World War II*. New York: Raintree, 2003.
Stanchack, John. *Eyewitness: Civil War*. New York: Dorling Kindersley, 2000.

Web sites

A History of National Security
    A historical exhibit from the Los Alamos National Laboratory.
    http://www.lanl.gov/history/index.shtml
How Stuff Works: How Radar Works
    An explanation of the science and technology behind radar.
    http://electronics.howstuffworks.com/radar.htm
The Radar Pages
    A history of British radar and Robert Watson-Watt.
    http://www.radarpages.co.uk
REME Museum of Technology
    A museum of military technology based in England.
    http://www.rememuseum.org.uk
Washington Museum of Military Technology
    A military museum based in Tacoma, Washington.
    http://www.wammt.org
World War II: How War Impacted Technology; How Technology Impacted War
    An online exhibition from the Institute of Electrical and Electronics Engineers.
    http://www.ieee-virtual museum.org/exhibit/index.php

OPTICS AND VISION

Books

Arnold, Nick. *Frightening Light and Sounds Dreadful*. New York: Scholastic Hippo, 2001.

Bromberg, Joan Lisa. *The Laser in America, 1950–1970*. Cambridge, MA: MIT Press, 1991.

Burnie, David. *Eyewitness: Light*. New York: Dorling Kindersley, 1998.

Cobb, Beckie, and Josh Cobb. *Light Action!: Amazing Experiments with Optics*. Bellingham, WA: International Society for Optics Engineering, 2005.

Hecht, Jeff. *City of Light: The Story of Fiber Optics*. New York: Oxford University Press, 1999.

Muirden, James. *How to Use an Astronomical Telescope*. New York: Fireside, 2005.

Zirker, J. B. *An Acre of Glass: A History and Forecast of the Telescope*. Baltimore, MD: Johns Hopkins University Press, 2005.

Web sites

Molecular Expressions: Science, Optics, and You
Background, history, and activities on the science of optics.
http://micro.magnet.fsu.edu/optics/index.html

Optics for Kids
Games, experiments, and lesson plans from the Optical Society of America.
http://www.opticsforkids.com

Telescope
A guide to different optical telescopes from the Institute and Museum of the History of Science.
http://brunelleschi.imss.fi.it/museum/esim.asp?c=200601

SCIENCE, TECHNOLOGY, AND MATHEMATICS

Books

Bridgman, Roger. *Eyewitness: Technology*. New York: Dorling Kindersley, 1998.

Bunch, Bryan. *The History of Science and Technology: A Browser's Guide to the Great Discoveries, Inventions, and the People Who Made Them from the Dawn of Time to Today*. New York: Houghton Mifflin, 2004.

Gribbin, John. *Science: A History: 1534–2001*. New York: Penguin, 2003.

Landes, David S. *Revolution in Time: Clocks and the Making of the Modern World*. Cambridge, MA: Belknap, 1983. Rev. and enlarged edition. Cambridge, MA: Harvard University Press, 2000.

Newmark, Ann. *Eyewitness: Chemistry*. New York: Dorling Kindersley, 2005.

Parker, Steve. *Eyewitness: Electricity*. New York: Dorling Kindersley, 2005.

Wilford, John Noble. *The Mapmakers*. Rev. ed. New York: Vintage, 2000.

Woodford, Chris. *Gravity: Routes of Science*. Farmington Hills, MI: Blackbirch, 2004.

Web sites

Chemical Achievers
    Biographies of chemical pioneers from the Chemical Heritage Foundation.
    http://www.chemheritage.org/classroom/chemach/index.html
Chemical Heritage Foundation: Classroom Resources
    Biographies, activities, and stories of science from the Chemical Heritage Foundation.
    http://www.chemheritage.org/classroom/class.html
Latitude and Longitude
    NASA page explaining the ideas behind coordinates and how they form the basis
    of timekeeping and time zones.
    http://www-istp.gsfc.nasa.gov/stargaze/slatlong.htm
Map Projections
    An exhibit on various kinds of projections from the U.S. Geological Survey.
    http://erg.usgs.gov/isb/pubs/MapProjections/projections.html
New York Hall of Science
    A hands-on science and technology center in Queens, New York.
    http://www.nyhallsci.org
Science Museum
    A museum of science and technology in London, England.
    http://www.sciencemuseum.org.uk

TRANSPORTATION

Books

Christopher, John. *Riding the Jetstream: The Story of Ballooning—From Montgolfier to Breitling*. London: John Murray, 2001.

Cummins, C. Lyle, Jr. *Internal Fire*. Rev. ed. Warrendale, PA: Society of Automotive Engineers, 1989.

Cunningham, Kevin. *The History of the Automobile*. Chanhassen, MN: Child's World, 2004.

Deford, Debra. *The Steam Engine*. New York: World Almanac, 2005.

Diesel, Eugene, et al. *From Engines to Autos: Five Pioneers in Engine Development*. Chicago: Regnery, 1960.

Editors of *Bicycling*. *The Noblest Invention: An Illustrated History of the Bicycle*. Emmaus, PA: Rodale, 2003.

Herbst, Judith. *The History of Transportation*. Breckenridge, CO: Twenty-First Century, 2005.

Herlihy, David V. *Bicycle: The History*. New Haven, CT: Yale University Press, 2004.

Jones, Glyn. *The Jet Pioneers: The Birth of Jet-Powered Flight*. London: Methuen, 1989.

Sutcliffe, Andrea. *Steam: The Untold Story of America's First Great Invention*. New York: Palgrave Macmillan, 2004.

Sutton, Richard. *Eyewitness: Car*. New York: Dorling Kindersley, 2005.

Watts, Steven. *The People's Tycoon: Henry Ford and the American Century*. New York: Knopf, 2005.

Williams, Harriet. *Road and Rail Transportation*. New York: Facts On File, 2004.

Woodford, Chris. *Ships and Submarines*. New York: Facts On File, 2004.

Web sites

Brief History of Rockets
> A chronology of rocketry compiled by NASA's space scientists.
> http://quest.arc.nasa.gov/space/teachers/rockets/history.html

Cars
> An exhibit on the history of the automobile by the Discovery Channel.
> http://www.discoverychannel.co.uk/cars/index.shtml

Chasing the Sun
> A history of early aviation by PBS.
> http://www.pbs.org/kcet/chasingthesun/index.html

Four-Stroke Internal Combustion Engine
> An animated exhibit of the working of the four-stroke engine by NASA.
> http://www.grc.nasa.gov/WWW/K-12/airplane/engopt.html

Hot Air Balloons
> An exhibit from the Exploratorium science museum.
> http://www.exploratorium.edu/ls/balloons/index.html

How Cars Work
> An explanation of car technology from Howstuffworks Web site.
> http://auto.howstuffworks.com/car.htm

How Does an Internal Combustion Engine Work?
> Animated artworks explaining the different stages of an engine's operation.
> http://www.keveney.com/otto.html

How Things Fly
> An interactive simulation about the basic principles of flight from NASA.
> http://www.aero.hq.nasa.gov/edu/simulation.html

Museum of Automobile History
> The world's largest collection of car history exhibits.
> http://www.themuseumofautomobilehistory.com

Pedaling History Bicycle Museum
> The Web site of the museum based in Orchard Park, New York.
> http://www.pedalinghistory.com

Smithsonian National Air and Space Museum
> Many exhibits about the history of flight.
> http://www.nasm.si.edu

Steamboats.com
> A page providing links to photos, memorabilia, articles, and research about steamboats.
> http://www.steamboats.com

Steam Engine Library
A collection of historical documents relating to the history of the steam engine.
http://www.history.rochester.edu/steam

The Wheelmen
The Web site of an organization dedicated to preserving early bicycles.
http://www.thewheelmen.org/

NAME INDEX

A **boldface** number preceding a colon indicates the volume. Numbers in *italic* indicate illustrations or other graphics. Numbers entirely in **boldface** refer to a full article.

Boot, Henry, **5**:1440

Borg, Anita, **2**:343

Borlaug, Norman, **1**:139–146; **2**:584, 587

Born, Max, **2**:558

Bourseul, Charles, **1**:80

Bouton, Charles-Marie, **2**:392, 398

Boyer, Herbert, **1**:147–156; **2**:587; **3**:786, 788

Boyer, Joseph, **1**:213, 216

Boyle, Robert, **3**:817, 820

Braille, Louis, **1**:157–164; **2**:327, 333, 342; **4**:1183; **5**:1332, 1493

Braithwaite, John, **2**:520–521, 522

Brandenberger, Jacques, **1**:16, **165–170**; **3**:829–830, 835

Brattain, Walter, **1**:57–68; **2**:341, 372; **3**:829, 836, 860; **4**:967, 969, 1163

Braun, Wernher von, **3**:707–708; **4**:1073

Brayton, A. P., **4**:1240

Breed, Allen, **1**:23, **171–176**; **5**:1413

Breed, Johnnie, **1**:175

Breguet, Louis, **5**:1339, 1341

Breit, Gregory, **5**:1437

Bridgwood, Charlotte, **1**:24

Brin, Sergey, **2**:373

Brown, Lesley, **2**:480–481

Brown, Louise, **2**:475, 480–481, *482*; **3**:788, 789

Brunel, Isambard Kingdom, **1**:19, **177–184**, 192; **2**:340, 485; **5**:1407, 1409, 1491

Brunel, Marc Isambard, **1**:19, 178

Brunelleschi, Filippo, **1**:198; **4**:1221

Bruno, Giordano, **3**:667

Budding, Edwin Beard, **1**:185–190

Burbank, Luther, **1**:203–210; **2**:584, 586

Burke, James, **2**:470

Burnham, Daniel H., **2**:571, 572, 621

Burr, John Albert, **1**:188, 189

Burroughs, William Seward, **1**:211–218; **2**:338, 341

Burroughs, William Seward, II, **1**:212

Burt, William, **5**:1332

Busch, Adolphus, **2**:438

Bush, Vannevar, **2**:339, 342

Bushnell, David, **1**:219–226; **4**:1076, 1077

Bushnell, Ezra, **1**:24, 220, 221

Bushnell, Nolan, **1**:227–234; **2**:506, 508

Butters, George, **1**:298, 300

Butterworth, William, **2**:423

Byron, Augusta Ada, **1**:43–44; **2**:343

Byron, Lord, **1**:43; **2**:342

Cai Lun, **2**:326, 333

Calley, John, **4**:1140, 1142, 1143, 1148

Campbell, Helen, **4**:*1124*

Campbell, Keith, **1**:235–242; **2**:587

Candler, Asa G., **4**:1246–1248

Carey, Frank, Sr., **5**:*1345*

Carlson, Chester, **1**:243–250; **2**:327, 335, 344, 374; **3**:860; **5**:1492–1493

Carnegie, Andrew, **4**:1175

Carnot, Nicholas Sadi, **2**:433

Carothers, Isobel, **1**:257

Carothers, Wallace, **1**:195, 202, **251–258**, 292–293; **2**:371, 513; **3**:722, 728, 787, 803, 804, 833, 835, 860

Carrier, Willis, **1**:197, 201, 202, **259–266**; **2**:516; **3**:829, 830, 835

Carson, Johnny, **3**:897

Carson, Rachel, **2**:513, 583

Cartwright, Edmund, **1**:287, 292

Carver, George Washington, **1**:267–276; **2**:584

Caselli, Giovanni, **1**:277–282; **2**:370

Cayley, George, **5**:1339

Celsius, Anders, **2**:534, 536

Chamberland, Charles, **1**:18; **4**:1214

Chandler, C. F., **1**:50–51

Chanel, Coco, **1**:54

Chanute, Octave, **5**:1479, 1482–1483, 1487

Charch, William Hale, **1**:168, 170; **4**:982

Chardonnet, Hilaire de, **1**:292, 293

Christie, Samuel Hunter, **5**:1446

Christina of Lorraine, **3**:670

Churchill, Winston, **4**:1077

Cicero, **1**:40

Cierva, Juan de la, **5**:1341

Clement, Joseph, **1**:43

Clerk, Douglas, **2**:439

Clinton, Bill, **2**:498

Cochran, Josephine, **1**:295–302; **3**:830, 832

Cochran, William, **1**:296, 298, 299

Coffman, Don, **1**:254, 256

Cohen, Stanley Norman, **1**:147–156; **2**:587; **3**:786, 788

Cole, Nat King, **4**:1233

Cole, Romaine, **3**:754

Coleman, Ornette, **5**:1306

Coleridge, Samuel Taylor, **4**:1118

Coles, Cowper, **2**:523

Colladon, Daniel, **3**:925

Collins, Arnold, **1**:253

Colt, Elizabeth Hart, **1**:309, 310

Colt, Samuel, **1**:303–312; **4**:1072, 1078

Coltrane, John, **5**:1306

Columbus, Christopher, **5**:1406

Compton, Arthur, **2**:563

Conant, James, **2**:563

Cook, James, **3**:777–778

Cooke, Charles, **2**:333

Cooke, William, **2**:327; **4**:1120; **5**:1446, 1447, 1448

Cooley, LeRoy, **3**:901, 902, 903

Coolidge, Calvin, **1**:273–274, 281

Cooper, Leon, **1**:65

Cooper, Martin, **2**:332, 336, **355–360**, 372

Copernicus, Nicolaus, **3**:666, 667; **4**:1150; **5**:1320, 1323

Corbusier, Le, **1**:191

Cormack, Allan, **2**:361–368; **3**:784, 789

Cornely, Paul D., **2**:445

Shockley, William, **1:**57–68;
2:341, 342, 372, 373, 374;
3:829, 836; **4:**967, 969, 1163

Sholes, Christopher Latham,
2:327, 334; **3:**793, 829,
834; **4:**1225; **5:**1329–1336

Siemens, William, and Frederick
Siemens, **1:***111*, 114

Sikorsky, Igor, **4:**1078;
5:1337–1348, 1397, 1413

Silbermann, Gottfried,
2:388–389

Silver, Spence, **1:**16, 17, 20;
2:372

Singa, Pradjeep, **1:***146*

Singer, Isaac Merrit, **1:**291;
3:841, 842, 843, 847

Skoll, Jeffrey, **4:**1173, 1175

Sloan, Alfred P., Jr., **3:**948

Smiles, Samuel, **4:***1147*

Smith, Oberlin, **2:**329

Soulé, Samuel, **5:**1330, 1332,
1333, 1335, 1336

Sousa, John Philip, **5:**1306

Spallanzani, Lazzaro, **1:**29

Spencer, Percy, **1:**15, 16; **3:**833

Springsteen, Bruce, **4:**1168

Stakman, Elvin Charles, **1:**140

Stalin, Joseph, **1:**274

Stanford, Leland, **4:**1175

Stanhope, earl of, **3:**646, 650

Stearns, J. B., **4:**1122

Stephenson, George, **2:**485, 489,
521; **3:**802, 804;
5:1349–1358, 1409

Stephenson, Robert, **2:**521;
5:1350, 1353, 1355, 1356

Steptoe, Patrick, **2:**475–482;
3:788

Sternbach, Leo, **3:**780

Stevens, William, **1:**114

Stockton, Robert, **2:**521–522

Strassman, Fritz, **2:**560, 561

Strauss, Levi, **1:**291, 293; **3:**861;
5:1359–1366

Strowger, Almon, **1:**83, 85

Sturgeon, William, **2:**544–545;
3:834; **4:**1119; **5:**1323,
1367–1373

Sueltz, Pat, **2:**343

Sugarman, Tracy, **4:***1050*

Sullivan, William, **2:**575

Swanson, Robert, **1:**152, 154

Symington, William, **3:**651

Szilard, Leo, **2:**564

Talbot, William Henry Fox,
2:333, 397, 399–400, 401,
501, 502, 506

Tanenbaum, Andrew,
5:1400–1401

Taylor, Zachary, **1:**308

Teetor, Ralph, **4:**1183

Telkes, Maria, **1:**201, 202; **2:**490,
517; **5:**1375–1380

Teller, Edward, **2:**489–490;
4:1073, 1075, 1080;
5:1381–1390

Tesla, Nikola, **2:**469; **4:**1019;
5:1391–1398

Thales, **5:**1427

Thimonnier, Barthélemy, **1:**288;
3:839

Thompson, Benjamin, **3:**828

Thompson, LaMarcus Adna,
4:1082, 1084

Thomson, William (Lord
Kelvin), **1:**78

Thoreau, Henry David, **2:**509

Tolman, Justin, **2:**548, 552

Torvalds, Linus, **2:**345, 346,
507; **4:**1226–1227, 1228;
5:1399–1404,
1491, 1493

Toussaint, Jean-Joseph-Henri,
4:1214, 1216

Townes, Charles, **2:**370; **3:**783,
787; **4:**1079, 1080, 1182,
1184; **5:**1307–1316

Trevithick, Richard, **5:**1351,
1409, 1410

Truman, Harry, **5:**1387

Ts'ai Lun, **2:**326, 333

Tull, Jethro, **2:**578,
579–580, 584;
3:801, 803

Tupper, Earl, **2:**513; **3:**829, 830,
835; **5:**1415–1424

Turri, Pellegrino, **4:**1183; **5:**1332

Tuve, Merle, **5:**1437

Twain, Mark, **1:**15, 20; **4:**1228

Tyndall, John, **3:**925

Urban VIII, Pope, **3:**671

Vail, Alfred, **4:**1120, 1125

Vanderbank, John, **4:***1150*

Verne, Jules, **5:**1338

Vesalius, Andreas, **3:**781, 784

Victoria, Queen (Britain), **1:**78,
109–110; **2:**544; **4:**979,
1043

Virchow, Rudolf, **3:**781, 786

Vitruvius, **2:**484

Volta, Alessandro, **2:**486, 488,
539, 611; **5:**1323, 1325,
1425–1434

Vukic, Ivan Blaz Lupis,
2:523–524

Walker, Samuel, **1:**307

Wan Hu, **3:**707

Waring, Fred, **4:**1232

Warren, John Collins, **4:**1129

Washington, Booker T., **1:**269,
270

Washington, George, 656; **1:**222;
3:648

Waterman, Lewis, **1:**130; **2:**327,
334

Watson, James, **3:**785, 788

Watson, Thomas, **1:**80–81

Watson, Thomas, Sr., **2:**341

Watson-Watt, Robert, **4:**1077,
1079; **5:**1397, **1435–1442**

Watt, James, **3:**651, 802; **4:**1146,
1147; **5:**1351

Watts, Isaac, **2:**538

Wayne, John, **5:**1363

Wayss, Gustav Adolf, **4:**1090

Wedgwood, Thomas, **2:**501

Welles, Orson, **2:**380

Wells, H. G., **3:**699

Wells, Horace, **4:**1128–1129, 1133

Wesley, Pam, **4:**1172, 1174, 1175

West, Benjamin, **3:**646–648, 654

Westinghouse, George, **5:**1394,
1396

Wheatstone, Charles, **2:**327, 333,
500; **4:**1120; **5:**1443–1448,
1490

Wheatstone, William, **5:**1444,
1445

Wheeler, George A., **4:**1258,
1260

INVENTIONS INDEX

A **boldface** number preceding a colon indicates the volume. Numbers in *italic* indicate illustrations or other graphics.

genetic engineering, **1**:147–156; **3**:786
  cloned animals, **1**:235, 236–241
  cloned human insulin, **1**:152, 153–154; **3**:786
  crops, **1**:155–156; **2**:583, 587
  patents, **1**:153, 154–155; **4**:1227–1228
geodesic dome, **1**:198, 202; **2**:623, 627–631, 632
G-Force vacuum cleaner, **2**:451, 452
Gibson guitar, **4**:1233
glaucoma treatment, **3**:780, 905, 907, 908, 911; **4**:1106, 1107
Global Positioning System (GPS), **3**:675, 678–682, 774, 775; **4**:*1017*
Google, **2**:373, 374, 376; **4**:1174
Gore-Tex, **1**:293
gramophone, **1**:79; **2**:329, 334, 467, 501
grandfather clock, **3**:853, 855
graphical user interface, **2**:375
gunpowder, **3**:707, 798, 800, 802; **4**:1069–1070, 1071, 1077, 1158, 1159, 1160, 1162; **5**:1320
guns, **1**:303–312; **4**:1021–1027, 1069–1073, 1077, 1078, 1158, 1160; **5**:1449, 1455, 1457
  interchangeable parts, **1**:303, 309, 311; **2**:592, 595; **4**:1072; **5**:1455, 1457
  naval, **4**:1075
Gutenberg press, **2**:326, 327, 328, 333; **3**:737, 740–747, 800, 802, 863; **4**:1177; **5**:1320, 1323, 1330, 1332, 1335

hair-straightening cream, **4**:1103
harness, **3**:799, 802
harpsichord, **2**:385, 386, 388, 500
harvester, **5**:1487, 1491–1492
headphones, **2**:501; **4**:1115
hearing aids
  Hooke prototype, **3**:820
  transistor, **1**:62–63, 66

heart transplant, **3**:783, 788
heart valves, artificial, **4**:1251, 1256
heat-resistant fibers, **4**:983–984, 985
helicopter, **1**:79; **4**:*1077*, 1078; **5**:1337–1347, 1397
herbicides, **1**:143, 156; **2**:513, 583
high-fidelity stereo, **3**:717–718
hot-air balloon, **1**:*179*; **4**:1093–1100; **5**:1410, 1413, 1479, 1496, 1497
household inventions, **3**:827–836
  utilitarian design, **3**:933, 934
  See also specific items
HTML (HyperText Markup Language), **1**:100, 104
HTTP (HyperText Transfer Protocol), **1**:100, 102
hula hoop, **2**:504
human growth hormone, **1**:152
human IVF. *See* in vitro fertilization
hybrid cars, **4**:*1204*, 1205
hybridization, **1**:206, 207
hydraulic press, **1**:282
hydraulic ram, **4**:1098
hydroelectric power, **2**:490; **4**:1237–1242
hydrofoil, **1**:82; **5**:1492
hydrogen bomb, **2**:490, 568; **4**:1073, 1080; **5**:1381–1389
hypertext, **2**:496–497

IBOT wheelchair, **2**:351; **3**:788, 916–918, 920
Idaho potato, **1**:203, 205–206, 210
ignition system, **3**:943, 946–947, 948, 952; **5**:1413
IM (instant message), **2**:331, 335; **4**:1034
iMac (computer), **3**:890
iMode, **2**:343
incinerator filtration, **3**:733, 735
incubator, **2**:554
inhalation anesthetics, **4**:1132, 1133
Ink Development (later eShop), **4**:1172, 1174
ink powder (toner), **1**:246, 248

insulin, **1**:152, 153–154; **2**:587; **3**:786, 788
integrated circuit, **1**:67; **2**:341, 342, 344, 345, 372; **4**:965–974, 1225
interchangeable parts, **1**:303, 309, 311; **2**:592, 595; **4**:1072; **5**:1455, 1457
internal combustion engine, **1**:94; **2**:512–513, 581; **5**:1369, 1410–1412
  Daimler, **1**:90; **2**:405–406
  Diesel, **2**:433–440; **5**:1412
  first successful, **2**:433
  Otto, **1**:89, 90; **2**:405, 433–434, 439, 440, 486, 489; **4**:1191–1198; **5**:1411–1412
  See also diesel engine; gasoline engine
Internet, **1**:86; **2**:325, 328, 335, 343, 344, 346, 373, 375, 376, 517–518; **3**:803, 804, 829; **4**:1117, 1123, 1124; **5**:1408
  chat programs, **4**:1034–1035, 1037
  collaboration, **2**:346; **5**:1402
  e-mail, **1**:282, 292; **2**:335, 357
  fiber-optic cables, **3**:923, 925; **4**:1182
  instant messaging, **2**:331, 335; **4**:1034
  interactive video games, **1**:233–234
  Linux downloading, **5**:1399, 1401, 1402
  online auctions, **4**:1171–1176
  online community, **4**:1033–1034, 1036, 1037
  search engine, **2**:373, 374, 376; **4**:1174
  telephone wires, **2**:357
  wireless access, **2**:360, 544; **4**:1011, 1016, 1018
  See also World Wide Web
intravenous anesthetics, **4**:1132
intravenous drug infusion pump, **3**:788, 914–915; **4**:919

in vitro fertilization (IVF),
    2:475–482; **3**:788, 789
iPhone, **2**:335, 336
iPod, **2**:335, 346; **3**:889, 890, 891
ironclads, **2**:522–524; **4**:1076
iTunes, **3**:890
Jacquard loom, **1**:43, 287, 292;
    2:338, 341; **3**:865, 868–872
jeans. *See* Levi's blue jeans
jet engine, **5**:1408, 1413–1414,
    1459–1467
jet helicopter, **5**:1344
jet plane, **3**:804; **4**:1077, 1078,
    1079; **5**:1441, 1459–1467,
    1486–1487

Kapton, **4**:1071
K brick, **1**:194–195, 202; **3**:931,
    934–936
Kevlar, **1**:290, 293; **2**:351, 372;
    4:981–986, 1071, 1080
kidney dialysis machine,
    mobile, **3**:788, *915*, 916,
    919, 920, 921
kinematoscope, **2**:502;
    4:1002–1003
kinetograph, **2**:334, 471, 502;
    4:1002, 1003
kinetoscope, **2**:334, *470*, 471, 504
kinetoscope parlors,
    4:1002–1003, 1008
Kleenex, **2**:430; **4**:1165
knitting machine, **1**:285–286, 291
Kodak camera, **2**:333, 373,
    456–457, 458, 502,
    504, 506, 507; **4**:1180,
    1181, 1182
    Polaroid suit, **4**:1226

laparoscopy, **2**:477, 478, 479,
    480, 481
laptop computer, **2**:*371*;
    5:1403, 1431
laser cataract surgery, **1**:69–74;
    3:783, 789; **4**:1184; **5**:1493
laser printer, **2**:375
laser technology, **2**:328, 370;
    4:1079, 1080, 1177,
    1182, 1184; **5**:1307,
    1312–1315, 1397

compact discs, **5**:1287, 1288,
    1315
fiber-optic cable, **3**:926, 927
optical disc, **5**:1289, 1315
optical disc reuse,
    4:1202–1203
printer, **2**:375
surgery, **1**:69, 71–74; **3**:783,
    787; **5**:1493
latex, **3**:733
lathe, automatic, **4**:1200
Latho (sponge), **3**:793, 794
laughing gas, **4**:1128–1129, 1132
lawn mowers, **1**:185–190; **2**:424
    powered, **1**:188; **4**:1183
LED displays, **2**:370
lenses, **4**:996, 998, 1152
    bifocal, **2**:606, 608, 613;
        4:1178–1179
    compound, **4**:1179–1180,
        1181; **5**:1320–1321, 1324
    convex, **4**:994–995
Levi's blue jeans, **1**:291, 293;
    3:861; **5**:1359–1366
life jackets, **2**:352
lightbulb, **1**:20, 197; **2**:461, 465,
    468, 473, 488, 489; **4**:1021,
    1022, 1026; **5**:1372
    design experiments, **3**:799
lightning rod, **2**:609;
    5:1323, 1489
linotype machine, **3**:747
Linux (operating system), **2**:345,
    346, 507; **4**:1227;
    5:1399–1404, 1491
Liquid Paper, **3**:829, 836, 860;
    4:1135–1138
Log, The (electric guitar),
    4:1233, 1235
logarithms, **1**:43, 47
    table, **5**:1322, 1324
longitude, **3**:773, 774–777, 778;
    5:1407
long-playing record (LP), **2**:329,
    501, 505, 507; **3**:711,
    714–715, 717–718
lubricator. *See* automatic lubricator

magnetic resonance imaging. *See*
    MRI scanner

magnetic tape, **2**:330; **3**:692–694;
    4:1234
magnetophone, **2**:329, 335
masers, **5**:1312, 1313, 1314
mass production, **2**:504, 509,
    590–593, 595; **3**:802, 863;
    4:1042; **5**:1492
    Whitney rifle, **5**:1449, 1455,
        1457, 1458, 1492
    See also assembly line; inter-
        changeable parts
meat preservation, **3**:385, 759,
    760–762, 832
Mercator map projection, **4**:1055,
    1057–1060; **5**:1407, 1409
mercury thermometer, **2**:531–535,
    536; **5**:1323, 1325
microchip, **1**:282; **2**:338, 342,
    344, 345, 506; **3**:832–883,
    885, 886; **4**:965, 969,
    971–972, 1256
microphone, **2**:472; **5**:1443,
    1444–1445
microscope, **3**:781, 782, 784, 789,
    801, 802, 817, 818; **4**:*1178*,
    1179–1180, 1181, 1210;
    5:1320–1321, 1324
microwave oven, **1**:15, 16;
    3:832–833
mines, naval, **1**:221,
    222–224, 306
Minix (operating system),
    5:1400–1401
missiles, **1**:65, 282; **2**:554; **4**:971,
    1069, 1073; **5**:1388
    proximity fuse detonator,
        4:1077
    radar-guided, **5**:1440
    "smart," **4**:1079
Model T car, **1**:24; **2**:589,
    593–595, 596, 600, 601;
    5:1412, 1413
modem, **1**:85–86
Morse code, **1**:78; **2**:328,
    462, 463; **4**:1120, *1121*,
    1124, *1125*
motion pictures, **2**:333–334, 401,
    502–503, 505
    Edison inventions, **2**:369,
        470–471, 473, 502

motion pictures (cont.)
  Lumières' invention of,
    4:999–1010, 1181, 1182
motorcycle, **2**:403, 405, 513;
    4:1065; **5**:1412
  Honda, **3**:805, 807, 808–814,
movable type, **2**:326, 327, 333;
    3:739–740, 742, 746, 747;
    5:1330, 1332
movie theaters, **2**:334, 471, 507;
    4:999, 1006, 1009
MP3 music player, **2**:337, 357
MRI scanner, **2**:351, 409–418;
    3:785, 789
Nerf guns, **3**:897
NiMH (nickel metal hydride)
    battery, **4**:1203–1204,
    1205, 1206
NMR spectroscopy, **2**:*411*,
    413–415, 416, 417;
    5:1325, 1326
nuclear energy, **2**:483, 488–490,
    491; **5**:1388
  power plants, **1**:16; **2**:557,
    567–568; **5**:1325, 1326
nuclear magnetic resonance. *See*
    NMR spectroscopy
nuclear reactor, **2**:489, 490,
    557–568; **4**:1073, 1079
nuclear weapons, **1**:16;
    2:489–490; **4**:1069,
    1073–1075, 1080, 1111,
    1164; **5**:1325, 1326, *1388*,
    1389. *See also* atomic bomb;
    hydrogen bomb
nylon, **1**:290; **2**:504, 513; **3**:787,
    833; **4**:983
  development of, **1**:195,
    200, 202, 251, 253–258,
    292–293; **2**:371; **3**:835, 860
  manufacture of, **3**:803–804

odometer, **2**:608; **3**:820
ore concentration process,
    2:527–530

paddle-wheel boats, **3**:650–653,
    655–656
paper bag, flat-bottomed, **3**:829,
    830, 834; **4**:975–980
paper clip, **2**:354

parachute, **1**:257; **4**:1098
pasteurization, **1**:31; **2**:582, 586;
    3:782, 785, 799; **4**:1207,
    1209–1216
peach, freestone, **1**:203
peanut uses, **1**:270–274, 275;
    2:437, 485, 584
Pelton wheel, **4**:1238–1240, 1241
pendulum clock, **3**:774–776,
    852–856; **5**:1320
penicillin, **1**:18; **3**:780, 783,
    787; **4**:1209
periodic table, **5**:1324–1325,
    1324
pesticides, **1**:143, 145; **2**:511,
    581, 582, 583, 587
phonograph, **2**:369, 464–473,
    500–501, 506
photocopier, **1**:243–250; **2**:328,
    335, 344, 374, 375; **3**:860,
    861; **5**:1492–1493
photography, **2**:332, 333,
    334, 346; **4**:1000,
    1001, 1002, 1009, 1181,
    1182; **5**:1328
  cell phone, **2**:335, 357, 360, 399
  color, **4**:1008, 1009
  development process,
    2:455–456
  digital, **2**:335, 338, 357,
    459–460, 507
  Eastman improvements,
    2:453–460
  invention of, **2**:391–401,
    501–502, 506, 606; **4**:1180,
    1181; **5**:1325
  patent infringement, **4**:1226
  satellite, **3**:717
  three-dimensional, **4**:1008,
    1177, 1182
  See also cameras; motion
    pictures
photovoltaic cells, **1**:200
photovoltaic film, **4**:1199, 1202,
    1204
piano, **2**:385–390, 500, 505
pickle machine, **3**:720–721
plant varieties, **1**:203–210; **2**:582,
    584, 586; **5**:1323–1324
  wheat improvement,
    1:139–145; **2**:584, 585

plastics
  development of, **1**:49, 53–55,
    195, 200, 202, 255; **2**:513;
    3:371
  explosives, **4**:1162
  fad items, **2**:504
  photodegradable, **2**:517
  photographic uses, **1**:169;
    2:333, 502, 506
  polymerization, **3**:833
  from soybeans, **1**:273;
    2:369–370
  See also nylon; Teflon
Playstation 3 game console, **2**:508
plum breeding, **1**:203,
    207–208, 209
pneumatic tire, **4**:1068
polyester, **1**:293
polymers, **1**:195, 251, 254–256;
    4:1252, 1253
Pong (video game), **1**:227,
    228, 229, 230, 231, 233,
    234; **2**:506
Portland cement, **1**:194; **4**:1089
Post-it Notes, **1**:16, 17, 20; **2**:372
PowWow (online community),
    4:1033–1034, 1036, 1037
printing, **3**:737–747
  Babbage plates, **1**:45
  Braille machine, **1**:161, 162
  See also Gutenberg press; mov-
    able type
punch cards, **1**:287; **3**:865, 867,
    868, 869, 871, 872
  Babbage calculator, **1**:43–44
  loom instructions, **3**:338,
    341, *871*
PVC (polyvinylchloride), **1**:255;
    3:833

Quadricycle, **2**:591
QWERTY keyboard, **5**:1331,
    1333, *1334*, 1336

radar, **1**:15; **2**:365, 366,
    494, 496, 554; **3**:677,
    678, 833; **4**:1069, 1077,
    1079; **5**:1309, 1310,
    1391, 1397
  development of, **5**:1435–1442
radiation therapy, **2**:362–363

spice sterilization, **3**:760–761, 832

spinet, **2**:386

spinning jenny, **1**:286, 287, 292; **3**:*763, 765, 767–772, 803*

spinning mule, **1**:287, 292; **3**:*765, 770–771*

spinning wheel, **1**:285; **3**:*766, 767*

spread-spectrum technology, **2**:332, 335; **4**:987, 991, *992*

stainless steel, **1**:113

stain-resistant fabrics, **1**:166–167, 170, 283, 294

steamboat, **2**:340, 520; **3**:802; **4**:1147; **5**:1407, 1410, 1411
  inventors of, **3**:645, 650–656, 803
  oceangoing, **1**:177, *178*, 180–183
  saloon, **1**:115, 116, 118
  screw propeller, **2**:521–526

steam engine, **2**:485–489, 512, 520–521, 522, 524, 584; **4**:1193; **5**:1350, 1369
  high-pressure, **5**:1351, 1353, 1409, 1410
  invention of, **2**:433–434; **3**:801, 802, 803; **4**:1139–1148
  lubricating cup, **4**:1051, 1054

steam locomotive, **2**:485, 489, 512, *520*, 521; **3**:802, 804, 861; **4**:1147; **5**:1350, 1351–1357, 1369, 1407–1411

steam loom, **1**:287, 288, 291, 292

steam press, **3**:742

steam turbine, **2**:488, 489; **5**:1396

steel
  Bessemer process, **1**:107, 110–114, 183, 193, 201; **2**:580, 618; **5**:1407, 1489
  as building material, **1**:*117*, 118, 193, 196, 199, 200; **2**:615, 618, 621; **3**:862
  electromagnet, **5**:1370
  reinforced concrete, **4**:1089, 1091

strengthened, **2**:539, 545

U.S. production, **1**:*111*, 116

steel plow, **2**:419–424, 511, 580, 585

stereoscope, **5**:1446

stethoscope, **3**:781, 790

submarine, **2**:525, 526; **3**:648–649, 652, 654; **4**:1076, 1077, 1078
  combat, **1**:219–226
  detection technology, **2**:554
  diesel engine, **2**:438; **5**:1412
  nuclear, **2**:567

subway, **4**:1260; **5**:1474, 1475

sugarcane-juice extractor, **1**:109–110

supermarkets, **2**:585, 586; **5**:*1298*, 1299–1300
  frozen-food case, **1**:*125*, 126

Super Soaker, **2**:504; **3**:893–898

supersonic flight, **5**:1414

suspension bridge, modern, **4**:1263–1272; **5**:1409

synthetics, **2**:513, 545
  drugs, **3**:905, 907, 908–910, 911
  fibers, **1**:290, 291–293; **2**:351, 427; **3**:860; **4**:983
  hormones, **3**:909
  rubber, **1**:16, 17, 195, 251, 253, 256, 273, 292; **2**:473; **3**:722, 728
  See also Kevlar; nylon; plastics; polymers

tape players, **4**:1113, 1114–1115

tape recorders, **4**:1112, 1114, 1234: *See also* videocassette recorder; videotape recorder

tape recording, **2**:329–330, 335, 336; **3**:692–693
  reel-to-reel, **3**:1234

Teflon, **1**:293–294; **2**:352, 371; **4**:1071, 1251–1256

telegraph, **1**:77–78, 79, 80, 277, 279–281, 306, 308
  "chemical," **1**:278–279
  as communication, **2**:327, 328–329, 330, 331, 333, 334
  Edison improvements, **2**:462–463, 472, 473; **5**:1490

electric current, **5**:1392

electromagnetic, **2**:543

induction system, **5**:1472

invention of, **2**:398; **3**:804

Morse improvement, **4**:1013, 1117, 1119–1126; **5**:1446

repeating, **5**:1490

underwater cables, **1**:183; **2**:327, 331; **4**:1122

Volta concept, **5**:1428

Wheatstone inventions, **5**:1443, 1446, 1447, 1448

telephone, **1**:277; **2**:341, 346; **4**:1015, 1020, 1122
  as communication, **2**:328, 329, 330, 331, 334, 517
  Edison improvements, **2**:464, 473
  fiber optics, **3**:923, 926, 930; **4**:1182; **5**:1491
  invention of, **1**:77–85; **2**:356, 357, 359, 370; **3**:799, 804; **5**:1493
  long-distance, **2**:328, 335
  wireless, **2**:332
  See also cell phones

telescope, **1**:116; **3**:817; **4**:995, 1149, 1151–1156, 1177, *1178*, 1179, *1180*, 1181, 1320; **5**:1321, 1328
  first, **4**:994–995; **5**:1320, 1324
  Galileo improvements, **3**:663, 667–670, 672, 673, 801, 802; **4**:997–998; **5**:1320, 1321
  Huygens lens, **3**:849, 851–852, 856
  multiple inventors, **4**:996

teletype, **5**:1335

television, **2**:331, 332, 337, 346, 503, 505; **3**:712–714; **4**:1181
  digital, **2**:507; **4**:1016
  electromagnetism, **2**:544
  fiber optics, **3**:923, 925
  invention of, **1**:19; **2**:547–555
  set numbers, **2**:335, *555*
  transistors, **1**:62; **4**:1114
  See also color television

telex machines, **1**:282

Windows (operating system), **3**:863, 864; **5**:1399, 1401, 1402
   software, **2**:344, 497, 498
windshield wiper, **1**:21–26; **5**:1413
wing warping, **5**:1482–1483
wireless technology, **2**:332; **3**:803; **5**:1391, *1436*
   cell phones, **2**:356–357, 360
   digital, **4**:1016
   electric power, **5**:1396, 1398
   Internet access, **2**:360, 544; **4**:1011, 1016, 1018

Marconi experiments, **4**:1013, 1014, 1020
   remote control, **5**:1395
   spread spectrum, **4**:987, 991
   See also radio
wire rope, **4**:1266, 1268, 1270, 1271
word processing, **2**:496; **3**:829; **5**:1335, 1336
World Wide Web, **2**:336, 343, 344–345, 508; **3**:804, 889; **4**:1033, 1184, 1228

   functions of, **2**:507
   independence of, **2**:375
   invention of, **1**:100–106; **5**:1490
   screen-reader programs, **4**:1183
wrought iron, **1**:110, 113, 183, 193
xerography, **1**:248, 249

Yahoo!, **2**:373, 374, 376; **4**:1034, 1035, 1037, 1174

COMPREHENSIVE INDEX

A **boldface** number preceding a colon indicates the volume. Numbers in *italic* indicate illustrations or other graphics. Numbers entirely in **boldface** refer to a full article.

Apple II computer, **2**:342, 345;
3:883, 884, 886–887,
890–892
Appleton, Edward, **5**:1437
Appleton, Victor, **3**:884
Apted, Michael, **3**:662
Aqua-Lung, **2**:377–384; **5**:1328
Arago, François, **2**:396
Archer, Frederick Scott, **2**:397,
400
Archimedes of Syracuse,
1:33–40; **2**:578; **3**:798;
5:1318–1319, 1323
Archimedes screw, **1**:33, 34–35,
36, 38, 40; **2**:578; **3**:798;
5:1319
architecture. *See* buildings and
materials
Aristarchus, **3**:666
Aristotle, **3**:663–664, 665; **4**:1153
Arizona
air-conditioning, **1**:265, 266
Navajo weaver, **1**:*284*
population growth, **1**:266
Arkwright, Richard, **1**:286–287,
292; **3**:765, 768–770
armor. *See* body armor
Armstrong, Neil, **2**:*554*, 555;
3:704
ArrayCom, **2**:358, 360
*Arrival of a Train at Ciotat Station,
The* (film), **4**:1004, 1006
artificial body parts, **1**:138, 200;
2:430; **3**:787; **4**:1008
eye research, **4**:1183
Teflon use, **4**:1251, 1256
artillery, **4**:1069, 1070, 1073, 1164
artists
Fulton, Robert, **3**:646, 654
Morse, Samuel, **4**:1118–1119
Art Shokai, **3**:806, 811
Aspdin, Joseph, **1**:194
aspirin, **3**:780, 786
Assayer, The (Galileo), **3**:671, 673
assembly line, **2**:509; **5**:1412, 1454
moving, **2**:589, 592–593, 594
See also mass production
Asteroids (video game), **1**:231
Aston, William Henry, **4**:1260

Astrodome, **2**:630, 631, 632
astronomy, **5**:1323
Galileo, **3**:665–672, 673, 801
Huygens, **3**:849, 851–852,
856, 857
laser use, **5**:1315
See also telescope
astrophysics, **5**:1313
AT&T, **1**:83, 781
Atanasoff, John, **2**:339, 342
Atari, **1**:22, 227, 229, 231, 234;
2:506, 508; **3**:885, 890
Atkins, Chet, **4**:1235
atom, splitting of. *See* nuclear
reactor
atomic bomb, **1**:67; **2**:557,
564–565; **5**:1383–1386.
See also nuclear weapons
atomic energy. *See* nuclear energy
atomic medicine, **3**:784
atomic physics, **2**:558–559, 611;
4:1073, 1079
Aubrey, John, **3**:816
AuctionWeb. *See* eBay
autochrome, **4**:1008, 1009
autogyro, **5**:1341
automatic devices. *See* key word
automobiles, **3**:*755*, 863–864;
5:1358, 1412–1413
air bag, **1**:23, 171–176; **5**:1413
air-conditioning, **1**:197
alternative fuels, **2**:437;
4:1199, 1203–1206
assembly line, **2**:589, 592–593,
594
battery, **5**:1430
Benz, **1**:87–95; **2**:374, 485,
489; **5**:1412
contest-winning inventions,
2:349, 351
cruise control, **4**:1183
Daimler, **2**:403–408, 485, 489;
4:1194, 1198; **5**:1412
diesel engine, **2**:439, 483, 485;
3:951, 952; **5**:1412
Dymaxion Car, **2**:627, 631
electric, **4**:1199, 1204, 1205;
5:1414
emissions control, **3**:812–813

engine knock, **3**:949
engine microchips, **4**:972
first, **1**:91; **5**:1412
Ford, **1**:24; **2**:589, 593–596,
598, 600, 601; **5**:1412, 1413
gasoline engine, **1**:87, 90–96;
2:403, 405, 406–407, 433,
483, 489, 513; **5**:1369, 1412
GPS receivers, **3**:681–682
Honda, **3**:811–814
hybrids, **3**:813; **4**:*1204*, 1205
ignition system, **3**:943,
946–947, 948, 952; **5**:1413
inventors of, **1**:87–95; **5**:1412
Kettering improvements,
3:943–952; **5**:1412–1413
mass production, **2**:509;
3:863; **5**:1412, 1457, 1492
numbers (21st century), **1**:94
painting system, **3**:734
plastic parts, **1**:200
pneumatic tires, **4**:1068
racing, **2**:406, 407
radar uses, **5**:1442
safety, **1**:23
sales (1900 to 1929), **1**:94
solar-powered, **4**:1199; **5**:1414
steam engine, **4**:1148
traffic signals, **4**:1104–1106;
5:1413
windshield wipers, **1**:21–26;
5:1413
See also tires
Automotive Hall of Fame, **1**:176
AutoSyringe, **3**:788, 915, 919
axle, **4**:1193
Axlon Inc., **1**:233

Babbage, Benjamin, **1**:42, 43
Babbage, Charles, **1**:41–48, 287;
2:338, 341, 343; **3**:867,
871; **5**:1321, 1322
Babbage, Henry Provost, **1**:45
Babylonians, **2**:337, 341; **5**:1232,
1318
Bach, Johann Sebastian, **2**:389
Bacon, Francis, **1**:121
Bacon, Roger, **4**:1069–1070,
1177, 1181

chess, **1:**228, 229

Chester and Lester (recording), **4:**1235

Chevrolet (car), **1:**174

Chicago
first skyscraper, **1:**195–196
Great Fire of 1871, **1:**195; **2:**621
See also World Columbian Exposition

Chicago school (architecture), **2:**621

children
Brownie camera, **2:**457, *458*
fads, **2:**504
invention contests for, **2:**351–352, 353
inventions by, **5:**1489–1494
spherical car seat, **2:**349
See also toys; video games

China, **1:**192; **2:**337
camera obscura, **2:**333
cloth, **1:**283, 284, 291
compass, **5:**1320, 1407
eyeglasses, **4:**1177
flying top, **5:**1338
globalization, **5:**1408
gunpowder, **3:**798, 802; **4:**1069, 1077, 1158; **5:**1320
herbal remedies, **3:**780
magnetic compass, **3:**799, 802
printing, **2:**326, 333; **3:**739–740, *741*, 800; **5:**1320
rocket, **3:**707
smallpox immunity, **3:**876, 877
toilet paper, **3:**834
transportation, **5:**1406

Chinook helicopter, **5:**1343

Chips Deluxe, **3:**941

chlorine, **4:**1252

chlorine compounds, **2:**539, 544

chlorofluorocarbons (CFCs), **2:**516; **3:**830, 950; **4:**1252, 1254

chloroform, **4:***1131*, 1132

cholera, **3:**659, 782, 786, 788; **4:**1214

Christie, Samuel Hunter, **5:**1446

Christina of Lorraine, **3:**670

chronometer, **2:**347; **3:**773–778

chronoscope, **5:**1446

Chrysler Corporation, **1:**93, 174; **2:**403

Chuck E. Cheese restaurants, **1:**227, 231–232

Churchill, Winston, **4:**1077

Cicero, **1:**40

Cierva, Juan de la, **5:**1341

Cinématographe, **2:**334, 503; **4:**1003–1005, 1009
how it works, **4:**1004

Civic (Honda compact car), **3:**812–813

Civil Rights Act (1964), **2:**445

civil rights movement, **4:**1054

Civil War, U.S., **1:**288; **2:**522–523; **3:**801; **4:**1046, 1070, 1072, 1258
anesthesia, **4:***1131*, 1133
Colt firearms, **1:**310, *311*
hot-air balloons, **5:**1496, 1500
railroads, **5:**1356
sewing machine, **3:**843, 844
slavery, **5:**1457

clarinet, **5:**1302, 1303, 1304, 1306

classification systems, **5:**1323, 1324, 1325

clavichord, **2:**387

clay, **1:**192, 194, 195

clay tablets, **2:**325, 333

Clement, Joseph, **1:**43

Clerk, Douglas, **2:**439

Clermont (steamboat), **3:**652–653, 655; **5:**1407, 1410
maiden voyage, **3:**653

Cleveland Clinic Foundation, **5:**1376

Clifton Suspension Bridge, **1:**179

climate control, **1:**197

Clinton, Bill, **2:**498

Clippers (flying boats), **5:**1341

clocks
chronometer, **3:**774–778
electromagnetic, **5:**1446
See also pendulum clock; watches

cloning
animal, **1:**235–242; **2:**587

basic steps, **1:***151, 241*

development of, **1:**147–156; **3:**786, 788

drugs, **1:**147, 152, 153

cloth and apparel, **1:283–294**; **2:**338
fireproof, **4:**1071
Jacquard loom, **1:**43, 287, 292; **3:**865–872
jeans, **5:**1359–1366
manufacture, **1:**286–291, 292; **3:**765–772, 801–803; **5:**1369
natural vs. artificial fibers, **1:**290
nylon stockings, **1:**200, 257–258; **2:**504, 513
ready-to-wear, **1:**291
sewing machine, **3:**837–847
stain-resistance, **1:**166–167, 170, 283, 294
synthetics, **1:**251, 255–256, 283, 290–293; **3:**803
textiles overview, **1:**290
textiles research, **4:**983
Velcro, **2:**425–430
waterproof, **1:**166–170, 293; **3:**720
See also cotton; weaving

coal, **2:**434, 490, *491*; **4:**1159; **5:**1350–1352, 1355
electric power, **2:**468; **5:**1434
external combustion engine, **1:**89; **2:**433
steam engine, **2:**485, 512, 581, 584; **4:**1143, 1146, 1147; **5:**1407
See also mining

coaxial cable, **1:**85

COBOL (computer language), **3:**824, 825

Coca-Cola, **2:**585, 586
invention of, **4:**1243–1250
trademark, **4:**1222
worldwide sales, **4:**1248

Coca-Cola Company, **2:**349, 351; **4:**1245, 1246

cocaine, **4:**1248

Cochran, Josephine, **1:295–302**; **3:**830, 832

computers (cont.)
  start-up philanthropists,
    4:1175
  supermarket uses, **5**:1300
  transistors, **1**:57, 62, 63, 67;
    2:366
  user-friendly systems, **2**:344,
    495–498, 506; **3**:829, 836,
    887, 890–891
  video games, **1**:227–231
  virtual war, **4**:1079
  virus effects, **4**:*1034*
  wireless, **3**:803
  word processing, **2**:496; **3**:829;
    5:1335, 1336
  See also Internet; software;
    World Wide Web
Computer Space (video game),
    1:229, 231
Computer Virus Industry
    Association (CVIA), **4**:1032
Conant, James, **2**:563
concertina, **2**:500; **5**:1445, 1448
concrete, **1**:192, 194–195, 200;
    2:472; **3**:935
  domes, **1**:198–199
  invention of, **4**:1089
  reinforced, **1**:194, 201; **3**:935;
    4:1087–1092
  See also K brick
condensation, **1**:263
Coney Island, **1**:195
  electric railway, **5**:1472,
    1473–1474, 1475
  escalator, **4**:1257, 1259, 1260
  Ferris wheel, **2**:574
  roller coaster, **4**:1082, 1083,
    1085
Constitution, U.S., **2**:612, 613
  patent law, **4**:1222–1223
consumer rights, **1**:173
container shipping, **3**:804
contests, **2**:347–354
  list of, **2**:353
  science, **3**:917, 921
  sponsors, **2**:349–350
Continental Edison, **5**:1393
conveyor belt, **2**:509

Cook, James, **3**:777–778
Cooke, Charles, **2**:333
Cooke, William, **2**:327; **4**:1120;
    5:1446, 1447, 1448
cooking, **3**:828, 834, 836
  solar, **5**:1378, 1379
  Teflon cookware, **4**:1251,
    1252, 1254–1256
coolants, **3**:830, 832
Cooley, LeRoy, **3**:901, 902, 903
Coolidge, Calvin, **1**:273–274, 281
cooling devices, **3**:828–829. *See
    also* air-conditioning
Cooper, Leon, **1**:65
Cooper, Martin, **2**:332, 336,
    355–360, 372
Cooperative Wheat Research and
    Production Program
    (Mexico), **1**:141, 145
Copernicus, Nicolaus, **3**:666,
    667; **4**:1150; **5**:1320, 1323
copier. *See* photocopier
copper
  extraction of, **1**:193, 201
  plates, **2**:394
Copyflo, **1**:249
copyright, **4**:1222; **5**:1403, 1404
Corbusier, Le, **1**:191
cordite, **4**:1070
Cormack, Allan, **2**:361–368;
    3:784, 789
corn, genetically modified, **1**:155
Cornell University, **1**:260, 265;
    3:730–731, 932, 936
Cornely, Paul D., **2**:445
cornflakes, **2**:585, 586; **3**:937,
    939–941
Corning Glass Company, **2**:373;
    3:929
Cornu, Paul, **5**:1339
Cornwallis, Charles, **1**:224
corporate community, **4**:973–974
corporate invention, **2**:369–376;
    3:804, 834, 859, 860, 863
  biotechnology, **1**:152, 153–154
  contest sponsorship, **2**:349–350
  Edison's influence, **2**:474
  integrated circuit, **4**:969, 973

young inventors, **5**:1493
  See also Bell Laboratories;
    DuPont Company
correction fluids, **4**:1135–1138
cortisone, **3**:909, 910
Cosimo II (Tuscany), **3**:670
Cosmotheoros (Huygens), **3**:852,
    857
cottage industry, **1**:285;
    3:769–772
cotton, **1**:*283*, 291, 293, 294
  disease, **1**:269, 270
  genetically modified, **1**:155
  growing of, **1**:269, 270, 274,
    275, 288–289;
    5:1450–1451, 1456–1457
  manufacture of, **3**:764, 765,
    766, 767, 769–772
  mercerization, **1**:166
  textiles, **1**:284, 285, 290
  U.S production (1790 to
    1860), **1**:288
cotton gin, **2**:509; **3**:801, 802,
    803, 844; **4**:1072
  how it works, **5**:*1453*
  invention of, **2**:288–292;
    5:1449–1452, 1492
  principle of, **5**:1452
Cousteau, Jacques-Yves,
    2:377–384; **5**:1325,
    1326, 1327
Cousteau, Philippe, and Jean-
    Michel Cousteau, **2**:380
Cousteau/Amazon (documentary),
    2:380
Cousteau Odyssey (documentary),
    2:380, 382
Cousteau Society, **2**:383
cowcatcher, **1**:47
Cowell, Simon, **2**:349
cowpox, **3**:874, 878–879, 881
Cox, Palmer, **2**:458
Craftsman/NSTA Young
    Inventors Award Program,
    2:352
Cramer, Stuart, **1**:262
crane, **3**:798
crankshaft, **4**:1196; **5**:1369

C ration, **1**:29

creationism, **2**:417

Creative Commons, **4**:1174

credit card machines, **5**:1300

credit cards, **4**:1184

credit sales, **3**:945–946; **4**:1042

Cremer, Gerard de. *See* Mercator, Gerardus

Cremoni, Cesare, **3**:668

Crick, Francis, **3**:785, 788

Criley, Thomas, **2**:529

Crimean War, **4**:1158

Cristofori, Bartolomeo, **2**:385–390, 500, 505

critical mass, **5**:1384

Crompton, Samuel, **1**:287, 291, 292; **3**:765, 770

crop rotation, **1**:270

Crosby, Bing, **4**:1233

crossbreeding, **1**:206, 207

CRT (cathode-ray tube) displays, **2**:494–495

cryolite, **3**:752

Crystal Palace (N.Y.C.), **4**:1187, *1188*

crystals, **2**:542, 544; **3**:818; **4**:1201, 1202

CT scanner, **2**:361–368; **3**:784, 789
how it works, **2**:364
MRI scanner vs., **2**:409

Cuban missile crisis, **4**:991

cuckoo nesting habits, **3**:874, 875

Cugnot, Nicolas-Joseph, **1**:89; **5**:1409, 1410

Cullen, Michael, **5**:1299

Cummings, Alexander, **3**:830, 833–834

cured meat, **3**:761, *762*

Curie, Marie, **2**:469; **3**:785, 786

current. *See* electric current

Curtiss, Glenn, **5**:1485

Curtiss, Lawrence, **3**:928

Curtiss-Wright Corporation, **3**:707, 732

cutlery, **1**:113, 118

CVCC Honda engine, **3**:812–813, 814
how it works, **3**:813

"Cycle Valse" (song), **4**:*1063*

Cyclone (Cleveland roller coaster), **4**:1084

Cyclone (Coney Island roller coaster), **4**:1083, 1085

cyclonic vacuum cleaner, **2**:449–452, 488

Cyclonite (RDX), **4**:1070

cylinder, volume of, **1**:40

cytotoxic drugs, **3**:780

Dabney, Ted, **1**:229, 231

Daguerre, Louis, **2**:333, **391–402**, 501, 506; **4**:1180, 1181, 1226; **5**:1325

daguerreotype, **2**:333, 501, 506; **4**:1180
equipment, **2**:396
invention of, **2**:394–399

Daimler, Gottlieb, **1**:90, 91, 93, 94; **2**:403–408, 485, 489, 513; **3**:863; **4**:1194, 1195, 1196, 1198; **5**:1412

Daimler, Paul, **2**:405, 407, 408

Daimler-Benz AG, **1**:93; **2**:407, 408

DaimlerChrysler, **1**:93; **2**:403

Daimler Motor Company, **2**:403, 406, 407, 408

Damadian, Raymond, **2**:351, **409–418**; **3**:785, 789

dams, **4**:1241

Darby, Abraham, **1**:192, 200

Darby, Abraham III, **1**:192

Darwin, Charles, **1**:204–205

Davis, Jacob, **1**:291, 293; **3**:861; **5**:1360–1361, 1365, 1366

Davy, Humphry, **2**:501, 538–539, 541, 542, 544, 545, 582, 585; **5**:1433

Dayton Laboratories and Engineering Company. *See* Delco

DDT (pesticide), **2**:513, 582, 587

deafness, **1**:76, 77, 79, 85; **5**:1493
hearing aids, **1**:62–63, 66; **3**:820

Dean, James, **5**:1363

Dearborn (Mich.) Ford Museum, **2**:369

decimal numbers, **2**:338

Declaration of Independence, **2**:612, 613

Deep Dipper roller coaster, **4**:1083, 1084

deep-sea diving. *See* underwater exploration

Deere, Charles, **2**:422, 423, 424

Deere, John, **2**:373, **419–424**, 511, 580, 585

Deere and Company, **2**:423, 424

deflagration, **4**:1159

De Forest, Lee, **1**:60; **4**:967, 1018; **5**:1395

dehydrated foods, **1**:127

dehydration, **5**:1377

DEKA Research and Development Corp., **3**:915–916, 918, 919, 920

Delarouche, Paul, **2**:396

Delco, **3**:946–948, 951, 952

de Mestral, George, **1**:16, 19, 293; **2**:425–430; **3**:799

Demologos (warship), **3**:655–656

denim, **3**:861; **5**:1362, 1364. *See also* jeans

Densmore, James, **4**:1225; **5**:1332, 1333, 1336

dentistry, **4**:1128–1129, 1130, 1132

deoxyribonucleic acid. *See* DNA

department-store escalators, **4**:1259

DePauw University, **3**:906, 907, 908, 909

De revolutionibus orbium coelestium (Copernicus), **3**:666

desalination plants, **2**:515, *517*

Descartes, René, **3**:850; **4**:1150, 1153

DeSimone, Livio, **2**:372

desk fan, electric, **3**:828, 834

Destroyer (torpedo boat), **2**:522, 524–525, 526

detonator, **4**:1070, 1077, 1159

Detroit Automobile Company, **2**:592, 600

Detroit Lubricating Company, **4**:1053

electric current (cont.)
  induction communication,
    5:1471
  positive and negative charge,
    2:610, 611
  transistor, **1**:62, 68
electric fields, **2**:544
electric furnace, **1**:114
electric guitar, **2**:500; **3**:860;
    4:1229–1236; **5**:1490
electricity, **2**:327, 486–488,
    539–545; **5**:1372, 1427–1428,
    1430, 1432, 1434, 1445
  Birdseye patents, **1**:127
  dynamo, **2**:468, *541*
  Edison inventions, **1**:196–197;
    2:468–472, 548, 581–582, 611
  Edison-Telsa rivalry, **2**:469
  electronics vs., **4**:967
  Faraday theory, **2**:541–542, 545
  farm use, **2**:581–582, 586
  Franklin experiments, **2**:607,
    609–610, 611, 613; **5**:1323,
    1325, 1427
  gold and iron extraction, **1**:193
  household appliances, **1**:196;
    3:827, 832, 836
  hydropower, **2**:490; **4**:1237–1242
  ignition system, **5**:1413
  internal combustion, **2**:433, 439
  ionized gas-generated, **3**:731–732
  photoconductivity, **1**:245, 246
  as power source, **2**:483, 540
  romance of, **5**:1372
  semiconductor switches,
    4:1200, 1201, 1202, 1204
  static, **1**:245, 246, 247; **2**:486;
    5:1426, 1427, 1428
  Sturgeon experiments,
    5:1367–1372
  Tesla inventions,
    5:1391–1398
  wireless transmission, **5**:1396
  See also battery; electronics;
    generator; power plants;
    specific applications
electric light, **1**:20, 196–197;
    2:369, 461, 465, 468, 471,
    473, 488, 489

electric motor, **3**:833, 834;
    5:1323, 1369, 1370–1371,
    1392–1393
electric railroad, **5**:1357, 1409
  devices for, **5**:1469–1475
  third rail, **5**:1474
electric shaver, **3**:688
electric typewriter, **4**:1136–1137;
    5:1334–1335, 1336
electrochemistry, **2**:541–542
electrocution, **2**:469
electrodes, **3**:752
electrogasdynamics, **3**:729,
    731–734
Electrolux (company), **2**:452
electrolyis, **1**:193
electrolyte, **5**:1430–1431
electromagnet, **2**:542, *543*;
    5:1323, 1367–1372, 1471
electromagnetism, **2**:327; **3**:784;
    4:1012, 1013, 1016, 1018,
    1019
  clock, **5**:1446
  Faraday theory, **2**:540–545
  Maxwell equations, **2**:544;
    5:1372
  medical imaging, **2**:413, 414
  radio, **2**:503
  telegraph, **4**:1119–1120,
    1125–1126
  Tesla coil, **5**:1395
electron, **1**:19, 62; **2**:542, 559, 611
electronic, games, **2**:349, 506
Electronic Frontier Foundation
    Pioneer award, **4**:991
electronics, **2**:338, 339, 356, 547
  companies, **2**:373
  electricity vs., **4**:967
  first computers, **2**:340, 343
  frequency-hopping, **4**:991
  integrated circuit, **4**:968–974
  microelectronics, **1**:67; **2**:359;
    4:968, 974
  Sony products, **4**:1112–1116
  See also specific applications
electronic video recording
    (EVR), **3**:717, 718
electron microscope, **2**:554
electrons, **3**:752; **5**:1431

electrophorus, **5**:*1426*,
    1427–1428, 1433
Elements (Euclid), **1**:38
elevator, **1**:195, 196, 201; **2**:618;
    3:862
  safety, **4**:1185–1190
elevator cars, **5**:1470, 1471
Elijah McCoy Manufacturing
    Company, **4**:1053, 1054
Eli Lilly and Company, **1**:152;
    3:780
Elizabeth I (England), **1**:286,
    287–288
Elliott, Charles Loring, **2**: *521*;
    1:310
Ellison, Larry, **4**:1175
El Paso Power Company, **5**:1396
Elugelab Island, **5**:*1386*, 1387,
    1389
e-mail, **1**:292; **2**:335, 357
embryology, **1**:236–239
embryonic stem (ES) cells,
    1:236–237
Emerson, Ralph Waldo, **2**:584
emgines, internal combustion,
    2:403, 405, 406, 434–440
EMI (company), **2**:366, 367, 368
emigration, **1**:177
emissions, **1**:95; **2**:437, 517–518
Emmy award, **3**:696
Empire State Building, **2**:622;
    4:1188; **5**:1502
endoscope, **4**:1182
endotracheal tube, **4**:1133
energy and power, **2**:483–492
  alternative fuels, **3**:*812*;
    4:1199, 1203–1206
  battery, **5**:1430–1434
  commutator, **5**:1369
  environmental impact,
    2:511–518
  fossil fuels, **2**:437, 487, 490,
    491; **5**:1408
  renewable sources, **1**:201, 202;
    2:487, 490–491, 514, 517;
    3:659; **4**:972, 1202, 1203,
    1242
  See also heat theory; power
    plants; *specific sources*

Energy Conversion Devices, Inc., **4**:1202, 1204, 1205

Energy Innovations, **3**:732

enflurane, **4**:1132

Engel, Joel, **2**:359

Engelbart, Christina, **2**:498

Engelbart, Douglas, **2**:344, 351, **493–498**; **3**:829, 830, 836; **4**:1225

engines. *See specific types*

English Channel, **4**:1015, 1020

ENIAC computer, **2**:338, 339, 340, 342, 343, 345; **4**:967, 971

Enquire (computer program), **1**:99

Enrico Fermi Award, **2**:566

entertainment, **2**:499–508. *See also specific types*

Environmental Protection Agency, **1**:189; **2**:513, 516

environment and inventing, **2**:351, **509–518**, 582, 583

air purifier, **3**:729, 733, 735–736

alternative-fuel cars, **4**:1204–1205

biodiesel fuel, **2**:437

building materials, **1**:200–201

Cousteau projects, **2**:383

Fuller (R. B.) projects, **2**:625–632

Gadgil projects, **3**:657–662

Honda auto emissions controls, **3**:811–814

hydroelectric plants, **4**:1242

incineration cleaner, **3**:733, 735

ozone layer hole, **3**:830, 950

refrigerant CFCs, **4**:1252

renewable energy sources, **2**:487, 490–491

transportation, **5**:1408

enzyme, restriction, **1**:149–150

Equanil, **3**:780

Ericsson, John, **2**:519–526**; **4**:1076, 1078; **5**:1407, 1411

escalator, **1**:195, 196, 201; **3**:862; **4**:1257–1261

escapement (gear), **3**:673

Escherichia coli. See E. coli

Esslinger, Friedrich Wilhelm, **1**:90

Etak Inc., **1**:233

ether anesthesia, **4**:1127, 1128, 1129–1130, *1131*, 1132–1133

Ethyl gasoline, **3**:949

Euclid, **1**:34–35, 38

eugenics, **1**:79

Europa (moon of Jupiter), **3**:667

European Court of Human Rights, **2**:451–452

European Patent Office (EPO), **4**:1219, 1227

European Space Agency, **3**:856

evaporation, **2**:610; **5**:1377

Eversharp, **1**:133

Everson, Carrie, **1**:193, 201; **2**:527–530**

Everson, George, **2**:550

EVI program, **4**:1204–1205

evolution theory, **1**:19; **3**:818

EVR (electronic video recording), **3**:717, 718

Experiments and Observations on Electricity (Franklin), **2**:607

ExploraVision, **2**:353

explosives, **1**:221, 222–224; **4**:1069–1070, 1073, 1078

chemistry of, **4**:1159

See also dynamite, gunpowder; nuclear weapons

external combustion, **1**:89; **2**:433

eyeglasses

bifocals, **2**:606, 608, 613; **4**:1178–1179

concave/convex lenses, **4**:994

eyes. *See* blindness; optics and vision

eye surgery, **1**:69–74; **3**:783, 789; **4**:1184; **5**:1315, 1493

Facebook.com, **4**:1034

factories, **2**:509, 512, 513–514

fads, **2**:504

bicycle, **4**:1066, 1067

Fahrenheit, Daniel, **2**:531–536**; **3**:781, 784, 801, 803; **5**:1323, 1325

Fahrenheit temperature scale, **2**:534–536; **3**:781, 801

Celsius vs., **2**:*534*, 536

Fairchild Semiconductor, **1**:66; **4**:969, 971, 973, 974

fairgrounds, **2**:505, 506, 571, 575

See also amusement parks

Falk, Benjamin, **3**:684

falling objects, speed of, **3**:663

Famous Amos (brand), **3**:941

fan, electric, **3**:828, 834

Fanta (soft drink), **4**:1249

fantoscope, **4**:1003

Faraday, Michael, **2**:488, 489, **537–546**, 611; **3**:833, 834, 862–863; **4**:1119; **5**:1323, 1369, 1433, 1434

electromagnetic induction, **5**:1471

generator, **5**:1445, 1446

Wheatstone friendship, **5**:1444, 1445

farming. *See* food and agriculture

Farnsworth, Elma Gardner ("Pem"), **2**:550, 555

Farnsworth, Philo, **1**:19; **2**:504, 505, 507, **547–556**

Farnsworth Wood Products Company, **2**:553

fascism, **2**:560; **4**:1018

fast-food chains, **1**:231–232

"Fat Man" (atomic bomb), **2**:565

Fat Tuesdays (jazz club), **4**:1235

fax machine, **1**:85, 277–282; **2**:328, 357, 517

first public demonstration, **2**:370

how it works, **1**:281

Federal Communications Commission, **3**:714, 715–716

feeding apparatus, **1**:135, 136–137; **3**:787, 788

Fellowship Corporation, **3**:940

Felt, Dorr E., **1**:214

felting, **1**:290

Fender, Leo, **4**:1231, 1233

fermentation, **1**:30; **4**:1208, 1209–1212, *1213*, 1216

Fermi, Enrico, **1**:16, 59; **2**:489, 490, **557–568**; **4**:1073, 1079, 1163; **5**:1325, 1326, 1383, *1385*

Burbank friendship, **1**:208
Carver friendship, **1**:274
corporation, **2**:373
Edison friendship, **2**:469, 472, 591
research laboratory, **2**:369–370
tractor, **2**:511, 580, 581, 586,
    593, 596, 601, 602; **5**:1491,
    1492
Ford, Henry, II, **2**:601
Ford, John, **2**:447
Ford, Mary, **4**:1235
Ford cars, **5**:*1412*, 1487
Ford Foundation, **5**:1378
Ford Motor Company, **1**:174;
    2:376, 592, 598, 599, 601;
    5:1492
    geodesic dome commission,
    2:629
Fordson Model F tractor, **2**:580,
    581, 586, 593, 596, 601, 602
forensic science, **1**:137, 138
formaldehyde, **1**:53
Forsyth, Alexander, **4**:1071, 1077
Forth and Clyde Canal, **3**:651
fossil fuels, **2**:437, 487, 490, 491;
    5:1408
    components breakdown, **2**:*491*
    See also coal; petroleum
fossils, **3**:818
Foucault, Pierre, **1**:161
fountain pen, **1**:130, 131, 132,
    134; **2**:327, 334
4D Company, **2**:625, 631
four-stroke engine, **1**:89, 90;
    2:405, 433–434, 439, 440;
    4:1191, 1194–1198
    how it works, **2**:*433*; **4**:*1196*
Fracastoro, Girolamo, **4**:1211
Francis, Samuel, **5**:1332
Francis B. Ogden (steamboat),
    2:521, 522
Franco-Prussian War, **4**:1065;
    5:1496–1497
Frankenstein (Shelley), **5**:1372
Franklin, Benjamin, **2**:486,
    603–614; **3**:654, 828, 833;
    5:1323, 1325, 1427, 1434
    bifocals, **2**:606, 608, 613;
    4:*1177*, 1178–1180

refusal to patent, **4**:1226
youthful inventions, **5**:1489,
    1493
Franklin, James, **2**:605
Franklin stove, **2**:605–606, 608,
    612, 613; **3**:828, 833
Frawley, Patrick, Jr., **1**:134
freeware, **5**:1402, 1403
freezers, **3**:830, 832
    display case, **1**:*125*, 126
freezing point, **2**:535
French Academy of Sciences,
    2:395–396, 398, 399;
    4:1130, 1133
French and Indian War, **2**:612
French Revolution, **3**:648, 867,
    870; **4**:1097
French Wine Coca,
    4:1244–1245, 1246
frequency-hopping technology,
    4:987–992
Fresca (soft drink), **4**:1249
Frigidaire Company, **3**:830, 950
Frisbee, **2**:504
    as generic term, **4**:1165
Frisbie Pie Company, **2**:504
Frisius, Gemma, **4**:1056
Froment, Paul Gustave, **1**:279, 280
Frosted Foods Company, **1**:124, 125
frozen foods, **1**:119–128; **2**:*582*,
    586; **3**:832, 835
    flash-freezing process, **1**:*124*
Fry, Arthur, **1**:16, 17; **2**:372
Fry, Jeremy, **2**:450, 452
fuel cells, **3**:893, 897
Fujisawa, Takeo, **3**:807, 808, 809,
    811, 814
Fuller, George, **1**:196;
    2:**615–622**; **3**:862
Fuller, Margaret, **2**:624
Fuller, Melville, **2**:624
Fuller, Ray, **3**:780, 788
Fuller, R. Buckminster, **1**:198,
    202; **2**:**623–632**
Fulton, Robert, **2**:485; **3**:**645–656**,
    803; **5**:1407, 1410
fur industry, **1**:121–122
fusion reaction, **5**:1383, 1384, 1387
Fust, Johann, **3**:744

Future City Competition, **2**:353
futurism, **1**:293

Gabor, Dennis, **4**:1182, 1184
Gadgil, Ashok, **3**:**657–662**, 788–789
Gagarin, Yury, **3**:704
Gagnan, Émile, 1327; **2**:377, 379,
    381; **5**:1325
Gale, Leonard, **4**:1120, 1125
Galen, **3**:779, 781, 784
Galilei, Galileo, **2**:533;
    3:**663–674**, 801, 802, 853,
    855; **4**:1150; **5**:1324
    telescope, **4**:997–998, 1179;
    5:1320, 1321
Galilei, Vincenzo, **3**:664
Galileo probe, **3**:894
Galileo spacecraft, **5**:*1321*
galleon (ship), **4**:1075–1076
galley (ship), **4**:1075, 1077
Galvani, Luigi, **2**:486, 488, 611;
    5:1372, 1428–1429, 1430
Gama, Vasco da, **5**:1406
games. *See* video games
gamma rays, **2**:367; **4**:998
G.A. Morgan Hair Refining
    Company, **4**:1103, 1106
Gandhi, Mahatma, **1**:274
gangsters, **4**:1023
Ganymede (moon of Jupiter), **3**:667
Garis, John, **1**:296
Garis-Cochran Dish Washing
    Machine, **1**:297–301
gas lamps, **2**:392
gas mask, **4**:1101, 1103–1104,
    1105; **5**:1417
    eyepieces, **1**:168, 170
Gasmotoren Fabrik Mannheim,
    1:90
gasoline, **2**:437, 439, 483; **4**:1192
    combustion, **5**:*1411*
    leaded, **3**:949, 952
gasoline engine, **2**:440, 472, 487;
    3:801
    airplane, **2**:485, 486, 513, 581;
    3:862; **5**:1477, 1481, 1484
    automobile, **1**:87, 90–96;
    2:403–407, 433, 483, 489,
    513; **5**:1369, 1412

gasoline engine *(cont.)*
> development of, **2**:512–513, 581; **5**:1411–1412
> farm machinery, **2**:581
> Honda fuel efficiency, **3**:812–813
> knock problem, **3**:949, 952
> as pollutant, **1**:95; **2**:437, 517–518
> *See also* internal combustion engine

gastroscope, **3**:788, 924, 928–929, 930; **4**:1182

Gates, Bill, **2**:344; **4**:1175

Gatling, Richard, **4**:1023, 1072

Gatling gun, **4**:1023, 1027, 1072

Gattoni, Giulio Cesare, **5**:1426

gauze bandage, **4**:1008

Gayetty, Joseph, **3**:834

gears, **1**:92; **2**:484; **3**:853; **5**:1319
> calculator, **5**:1322, 1490
> invention of, **3**:798

gelatin, **1**:31–32

Genentech, **1**:147, 152, 153–154

General Bakelite Corporation, **1**:54, 55

General Electric, **1**:15, 16, 17, 248; **2**:374; **3**:832, 834; **4**:1138; **5**:1286, 1287, 1475
> first color television set, **3**:717
> research laboratory, **2**:370–371

General Foods, **1**:124, 125; **3**:939

General Magic, **4**:1172

General Motors, **1**:174; **2**:600; **3**:948–952; **4**:1204–1205

General Seafoods Company, **1**:123, 124

generator, **2**:468, 488; **3**:862–863, 922; **5**:1371, 1392
> electrogasdynamics, **3**:731–732, 733
> invention of, **2**:537, 540, 541, 544, 545; **5**:1445, 1446

genetic engineering, **1**:147–156; **3**:786
> cloned animals, **1**:235, 236–241
> cloned human insulin, **1**:152, 153–154; **3**:786

crops, **1**:155–156; **2**:583, 587

patents, **1**:153, 154–155; **4**:1227–1228

genetics
> plant breeding, **1**:203–210; **5**:1323–1324, 1325
> *See also* DNA

Geocaching (GPS game), **3**:682

geocentric theory, **3**:666–667, 670, 671; **4**:997; **5**:1318

geodesic dome, **1**:198, 202; **2**:623, 627–631, 632
> how it works, **2**:628

Geodesics, Inc., **2**:629–630

geometry, **1**:38, 40; **5**:1318

George III (Britain), **3**:654, 777

George IV (Britain), **3**:880

George A. Fuller Company, **2**:615, 618, 621–622

George Washington Bridge, **4**:1271

George Washington Carver (film), **1**:273

George Washington Carver Foundation, **1**:274

George Washington Carver National Monument, **1**:274–275

geotextiles, **1**:293

geothermal energy, **2**:491

Gerlach, Walter, **3**:925

germanium, **1**:60, 61; **4**:968, 970

germ theory, **3**:782, 785; **4**:1207, 1209, 1211, 1214, 1216

Getting, Ivan I., **3**:675–682

G-Force vacuum cleaner, **2**:451, 452

Gibbons, Eric, **2**:350

Gibbs, James, **3**:843

Gibernau, Sete, **3**:808

Gibson guitar, **4**:1233

Gihon Foundation, **4**:1135, 1138

Gillette, King C., **3**:683–690, 833, 834, 860

Gillette Company, **3**:686–687, 688, 689; **4**:1138

Gillray, James, **3**:878

Ginger. *See* Segway

Ginsburg, Charles, **2**:329, 330, 335, 505; **3**:691–698, 829, 836

glass jars, **1**:29, 30, 31

Glauber's salt, **5**:1379

glaucoma, **4**:1106, 1107
> treatment, **3**:780, 905, 907, 908, 911

Glidden, Carlos, **5**:1330, 1332, 1333, 1335, 1336

Glidden Company, **3**:907, 909, 910

gliders, **5**:1413, 1477, 1479, 1481, 1482, 1484, 1487

Global Command and Control System, **4**:1079, 1080

globalization, **3**:804; **5**:1408

Global Positioning System (GPS), **3**:675, 678–682, 774, 775; **4**:1017
> how it works, **3**:679

Global System for Mobile Telecommunications, **2**:359

global warming, **2**:437, 487, 490, 514–515, 517; **5**:1408

Gloster E28/39 (jet test plane), **5**:1464, 1467

go (Japanese game), **1**:228, 229

Goddard, Robert H., **3**:699–710; **4**:1073, 1079, 1227; **5**:1325, 1413, 1414, 1491

Goddard Space Flight Center, **3**:709

Goizueta, Roberto, **4**:1249

gold
> extraction, **1**:193, 201
> measurement, **1**:36–37
> *See also* gold rush

Goldberg, Reuben, **2**:352, *354*

Golden Gate Bridge, **4**:1271

Golden Spike National Historic Site (Utah), **5**:1407

Goldmark, Karl, **3**:712

Goldmark, Peter, **2**:332, 335, 501, 505, 507; **3**:711–718, 830; **4**:1182

gold rush, **1**:291; **4**:1237, 1238–1240, 1241

Goldsmith, Michael, **2**:413

Goldstein, Adele, **2**:343

harness, **3**:799, 802

Harney, William, **1**:306

harpoon, **3**:797

harpsichord, **2**:385, 386, 388, 500

Harrar, George, **1**:141

Harrison, John, **2**:347; **3:773–778**

Harrods department store (London), **4**:1259

Hart, Elizabeth. *See* Colt, Elizabeth Hart

Harvard Business Review, **3**:861

Harvard Mark I (digital computer), **2**:339, 342, 343

Harvard Medical School, **2**:411; **4**:1128, 1129, 1130

Harvard University, **1**:58, 252; **2**:339, 371, 413, 624; **3**:676, 860, 906

  computers, **2**:339, 342, 343; **3**:822–823, 824

harvester, **5**:1487, 1491–1492

Harvest of the Years, The (Burbank), **1**:*208*

Hasbro, **3**:896, 897; **4**:1173

Haüy, Valentin, **1**:151, 159, 164

Haydn, Franz Josef, **2**:389

Hayward, Nathaniel, **3**:722

headphones, **2**:501; **4**:1115

health and medicine, **1**:18; **2**:351, 515; **3:779–790**

  anesthesia, **3**:780, 783, 785; **4**:1127–1133, 1127–1134

  biotechnology patents, **1**:153, 155; **4**:1227–1228

  blood banks, **2**:441–448; **3**:783, 787, 789

  CT scanner, **2**:361–368; **3**:784, 789

  fiber-optic instruments, **3**:928–929, 930; **4**:1182

  inventions, **3**:788, 801, 914–918

  laser uses, **4**:1184; **5**:1493

  Lumières' research, **4**:1008

  MRI scanner, **2**:409–418

  nuclear technology, **5**:1325, 1326

  racial segregation, **2**:445, 447

Teflon uses, **4**:1251, 1256

Velcro uses, **2**:430

in vitro fertilization, **2**:475–482

  See also diseases; drugs; germ theory; surgery; vaccine and vaccination

hearing aids

  Hooke prototype, **3**:820

  transistor, **1**:62–63, 66

heart disease, **3**:881

heart transplant, **3**:783, 788

heart valves, artificial, **4**:1251, 1256

heat, **4**:998

  air-conditioning, **1**:261–266

  canning process, **1**:29–32

  explosives, **4**:1159

  fermentation process, **1**:30

  furnace, **1**:110–118, *116*

  iron ore, **1**:110, 112–113, *114*

  pasteurization, **1**:31; **4**:1212

  plastics, **1**:53

heat engine, **2**:432, 512, 524

heat flow, **2**:610

heating systems, **1**:266; **3**:827–828, 833

  solar, **5**:1376–1377, *1378*, 1379–1380

  See also Franklin stove

heat lamp, **1**:127

heat-resistant fibers, **4**:983–984, 985

heat theory, **3**:818, 828; **5**:1323

Hebron, Charles, **2**:529

Heel, Abraham van, **3**:928

Hegel, Georg Wilhelm Friedrich, **4**:1264

Heisenberg, Werner, **5**:1382–1383

helicopter, **1**:79; **4**:*1077*, 1078; **5**:1337–1347, 1397

  how it works, **5**:1342–1343

  model, **5**:1478, 1479, 1491

heliocentric theory, **3**:666, 667, 668, 671, 673, 801; **4**:1179; **5**:1320, 1323

heliography, **2**:393

helium, **4**:1100; **5**:1500–1501

Helmholtz, Hermann, **1**:80

Henderson, Shelby, **3**:696

Henle, Friedrich, **4**:1211

Hennebique, François, **1**:194

Henry VIII (England), **4**:1221

Henry, Beulah, **3:791–796**

Henry, Joseph, **2**:543, 544–545; **4**:1119; **5**:1369

Henry Bessemer and Company Limited, **1**:111, 117, 118

Henry Ford Museum (Dearborn, Mich.), **2**:464, 596

Henry Umbrella and Parasol Company, **3**:793

hepatitis, **3**:788

Heracleides, **1**:34

herbal remedies, **3**:780

herbicides, **1**:143, 156; **2**:513, 583

Héroult, Paul, **1**:193; **3**:753

Herschel, Caroline, **4**:1180

Herschel, William, **4**:1180

Hertz, Heinrich, **4**:1012, 1019; **5**:1437

Hewlett, James Monroe, **2**:624

Hewlett-Packard, **2**:375; **3**:884, 886, 890

hexamethylenediamine, **1**:255

Hickman, Clarence, **4**:1073

hieroglyphs, **5**:1318

Hiero II (Syracuse), **1**:34, 35, 36, 37, 38–39

high-fidelity stereo, **3**:717–718

Higinbotham, William, **1**:230

Hill, Julian, **1**:253–254, 256

Hills, Amariah, **1**:187, 189

Hindenburg (dirigible), **5**:1500, *1501*

Hippocrates, **3**:779, 781, 784

Hippocratic oath, **3**:781

hip replacement, **3**:786

Hiroshima, atomic bombing of, **2**:565, 567; **4**:1080, 1111; **5**:1385–1386

Hirschowitz, Basil, **3**:928

His Master's Voice (Berraud), **2**:329

history of invention, **3:797–804**

  Edison's significance, **2**:473–474

  innovation, **3**:859–864

history of invention (*cont.*)
  mathematics, **1**:38–40;
    5:1318–1328
  myths, **3**:799
  patents, **4**:1221–1228
  science, **5**:1318–1328
  technology, **5**:1318–1328
  transportation, **5**:1405–1414
  weapons, **4**:1069–1080
  young inventors, **5**:1489–1494
Hitler, Adolf, **2**:564, 597–598;
    4:988, 1018, 1073, 1079;
    5:1383, 1437, 1463
HIV. *See* AIDS/HIV
H.J. Heinz Corporation, **5**:1299
Hodgeson, Peter, **1**:17
hoe, **2**:580
Hoe, Richard March, **3**:747
Hoechst (company), **2**:369
Hoff, Marcian Edward ("Ted"),
    2:342, 344; **3**:832,
    885; **4**:971
Hoffmann, Felix, **3**:780, 786
Holabird, William, **2**:621
Holland, John, **4**:1076, 1078
Hollerith, Herman, **2**:338–339,
    342
Hollywood, **2**:503; **4**:1008
Holmes, Oliver Wendell, **4**:1133
Holocaust, **2**:560
hologram, **4**:1177, 1182, 1184
Homebrew Computer Club, **3**:886
Home Insurance Building, **1**:196
Honda, Soichiro, **3:805–814**;
    5:1412
Hondo Motor Company,
    3:807–814
Hooke, Robert, **1**:292; **3**:781,
    784, 801, 802, **815–820**;
    4:1142, 1179–1180, 1181;
    5:1320, 1324
Hooke's Law, **3**:817
Hoover (company), **2**:452
Hoover, Herbert, **2**:469; **3**:940
Hopkins, Harold, **3**:924, 925, 928
Hopper, Grace Murray,
    2:341–342, 343, 344;
    3:821–826
hormones, **1**:152, 153; **3**:908, 909

horses, **4**:1075, 1077; **5**:1320
horticulture. *See* food and agriculture; plants
Hoskyns, John, **3**:816
hot-air balloon, **1**:*179*;
    4:1093–1100; **5**:1410, 1413,
    1479, 1496, 1497
  world records, **4**:*1099*
Hounsfield, Godfrey, **2:361–368**;
    3:784, 789
household inventions, **3:827–836**
  air conditioners, **1**:265
  air purifiers, **3**:735–736
  aluminum pots, **3**:755
  common items, **3**:830
  cyclonic vacuum cleaners,
    2:449–452
  dishwasher, **1**:295–302
  electric light, **1**:197
  electric-powered, **2**:488
  Franklin stove, **2**:605–606,
    612, 613
  lawn mower, **1**:185–190; **4**:1183
  microwave oven, **1**:15, 16, 67;
    3:832–833
  plastic items, **1**:53, 54, 55, 169
  sewing machine, **3**:837–844
  stainless steel items, **1**:113
  television sets, **2**:335, 555
  Tupperware, **5**:1415–1423
  utilitarian design, **3**:933, 934
houses
  first solar-powered, **2**:490, 517
  Fuller (R.B.) designs,
    2:625–627
  solar, **5**:1376–1377, 1378,
    1379–1380
Howard University, **2**:442, 444,
    446; **3**:906, 907
Howard University School of
    Medicine, **1**:70–71, 72, 73
Howe, Elias, **1**:291; **3:837–844**,
    847, 862; **4**:1225
"How High the Moon" (song),
    4:1235
HTML (HyperText Markup
    Language), **1**:100, 104
HTTP (HyperText Transfer
    Protocol), **1**:100, 102

Huang Ti, **1**:783
Hubbard, Gardiner Green, **1**:79
Hubbard, Mabel. *See* Bell, Mabel
    Hubbard
Hubert, Brian, **2**:*347*
Hudson River, **3**:650, 652–653
  map of *Clermont*'s maiden voyage, **3**:*653*
Hughes Laboratory, **5**:1312
hula hoops, **2**:504
Hulsmeyer, Christian,
    5:1436–1437
human cloning, **1**:240–241
Human Drift, The (Gillette),
    3:686, 688
Human Genome Project, **3**:786
human growth hormone, **1**:152
human IVF. *See* in vitro
    fertilization
humidity, **1**:260, 261, 262
  lowering of, **1**:263, 266
hunger, **1**:139, 141, 144, 145
Hunt, Alfred, **3**:754
Hunt, Walter, **3**:828, 839, 841,
    845–848
Hunter, John, **3**:874
Hunter College, **1**:70
  Hall of Fame, **1**:73
Hussey, Obed, **4**:1043
Huygens, Christian, **3:849–858**;
    5:1320, 1324, 1410
Huygens, Constantin, **3**:850
hybrid cars, **4**:*1204*
  sales, **4**:1205
hybridization, **1**:206, 207
Hyde, James, **2**:530
hydraulic press, **1**:282
hydraulic ram, **4**:1098
hydraulics, **1**:34
hydrocarbons, **2**:437
hydrodome, **1**:82, 85
hydroelectric power, **2**:490;
    4:1237–1242
  electromotor commutator,
    5:*1371*
hydrofoil, **1**:82; **5**:1492
hydrogen, **2**:437
  atoms, **2**:415
hydrogen balloon, **4**:1095, 1100

international transportation,
5:1408, 1500. *See also specific forms*
Internet, 1:86; 2:343, 344,
346; 3:803, 804, 829;
5:1408
antivirus software,
4:1031–1033, 1036–1037
billionaires, 4:1173, 1174, 1175
chat programs, 4:1034–1035,
1037
collaboration, 2:346; 5:1402
communication, 2:325, 328,
335, 517–518
corporations, 2:373, 375, 376
e-mail, 1:282, 292; 2:335, 357
as entertainment, 2:507
fiber-optic cables, 3:923, 925;
4:1182
hardware for the blind, 1:*163*
instant messaging, 2:331, 335;
4:1034
interactive video games,
1:233–234
Linux downloading, 5:1399,
1401, 1402
military, 4:1079
online auctions, 4:1171–1176
patents, 4:1228
patent tracking, 4:1224
PowWow online community,
4:1033–1034, 1036, 1037
search engine, 2:373, 374,
376; 4:1174
telegraph compared with,
4:1117, 1123, 1124
telephone wires, 2:357
wireless access, 2:360, 544;
4:1011, 1016, 1018
See also World Wide Web
Internet café, 1:*101*
Internet Explorer, 1:101, 102
intravenous anesthetics, 4:1132
intravenous drug infusion pump,
3:788, 914–915; 4:919
Introduction to Atomic Physics
(Fermi), 2:558, 559
Inuit, 4:1071
InvenTeams, 2:353

invention and innovation,
3:859–864
invention contests. *See* contests
Invention Quest, 2:353
Invent Now America (competition), 2:350, 353
in vitro fertilization (IVF),
2:475–482; 3:788, 789
Io (moon of Jupiter), 3:667
ionized gas, 3:731–732, 736
ionosphere, 4:1016; 5:1437
ions, 5:1431
Ioptics, 5:1290–1291, 1292
iPhone, 2:335, 336
iPod, 2:335, 346; 3:889, 890, 891
Iraq War, 4:1079, 1080
iron
buildings, 1:192–193, 198, 200, 201
cable suspension bridge,
4:1266, 1267
concrete reinforcement,
4:1088, 1089, 1090
domes, 1:198
forms of, 1:110, 113, 192–193
ships, 1:183; 2:522–524;
4:1076; 5:1407
See also steel
Ironbridge, 1:192, 200
ironclads (warships), 2:522–524;
4:1076
Iron Foundry and Mechanical
Workshop, 1:88, 92
Iroquois Nation, 3:876
irradiated food, 2:586
irrigation, 1:35, 145; 2:511–512, 578
Islamic world, 3:799; 5:1319
Isle of Man Tourist Trophy race,
3:808, 809
isochronicity, 3:853, 855
isoflurane, 4:1132
iSoldIt, 4:*1176*
Italian Space Agency, 3:856
ITT, 2:553–554
iTunes, 3:890
Ive, Jonathan, 3:889

Jackson, Charles T., 4:1119,
1120, 1122–1123, 1125,
1128, 1129, 1130, 1133

Jacobs, Eli, 5:1288
Jacquard, Joseph-Marie, 1:43,
287, 292; 2:338, 341;
3:865–872
Jacquard loom, 1:43, 287, 292;
2:338, 341; 3:865, 868–872
how it works, 3:869
Jansen, Hans, 4:995
Jansen, Zacharias, 4:995
Japan, 1:192; 2:343
electronics, 4:1112–1116
samurai armor, 4:1071
World War II, 4:1073, 1080,
1110–1111, 1111, 1164;
5:1385–1386, 1440
Java (computer language), 2:343,
345
Jaws (film), 1:229
jazz, 5:1301, 1303, 1305, 1306
jeans, 5:1359–1366
origin of word, 1362
Jefferson, Thomas, 1:224; 2:*613*;
3:880
Jellinek, Emil, 2:407, 408
Jellinek, Mercedes, 2:407, 408
Jenkins, Charles Francis, 2:553;
3:925
Jenner, Edward, 3:782, 785,
873–882
Jenney, William Le Baron, 1:196;
3:862
jet engine, 5:1408, 1413–1414
how it works, 5:*1461*
invention of, 5:1459–1467
jet helicopter, 5:1344
jet plane, 3:804; 4:1077, 1078,
1079; 5:1441, 1459–1467,
1486–1487
Jet Propulsion Laboratory,
3:731–732, 893, 896, 897
J.L. Hudson department store, 1:264
Jobs, Steve, 2:334, 336, 342, 344,
345, 373, 374, 497, 506,
508; **3:883–892**, 918, 920;
5:1491, 1493
John A. Roebling Suspension
Bridge, 4:1269
Johnson & Johnson, 3:918
Johnson, Lonnie, 2:504; **3:893–898**

Johnson, Lyndon, **1**:173
Johnson Research &
    Development, **3**:893, 897
Joliot-Curie, Irène, and Frédéric
    Joliot-Curie, **2**:559, 560
Jones, Amanda, **3**:899–904
Jules Stein Eye Institute, **1**:71
Julian, Percy Lavon, **3**:780, 787,
    905–912
Julian Laboratories, **3**:909, 910
junction transistor, **1**:61
Jupiter (locomotive), **5**:*1407*
Jupiter (planet), **3**:894; **5**:*1321*
    moons of, **3**:*667*, 670, 672,
    673; **4**:997

Kámán, Theodor von, **5**:1341
Kamen, Bart, **3**:914–915
Kamen, Dean, **2**:351; **3**:788,
    913–922
Kao, Charles, **3**:929
Kapany, Narinder, **2**:328, 335,
    373; **3**:787, 788, **923–930**;
    4:1182; **5**:1491
Kapany, Rajinder, **3**:929
Kapton (light material), **4**:1071
Kaptron (company), **3**:925, 929
Kay, John, **1**:286, 287, 292
K brick, **1**:194–195, 202; **3**:931,
    934–936
Kearns, Robert, **1**:24
Keebler, **3**:941
Keichline, Anna, **1**:194–195,
    202; **3**:931–936
Keith, Sir William, **2**:605
Keller, Helen, **1**:78, 208
Kellogg, John Harvey, **2**:585,
    586; **3**:938, 939, 941
Kellogg, William Keith, **2**:585,
    586; **3**:937–942
Kellogg's Corn Flakes, **3**:937,
    939–940, 941, 942
Kelly, William, **1**:116
Kelvin, Lord (William
    Thomson), **1**:78
Kemper, Steve, **3**:918
Kennedy, John F., **2**:348, 384;
    3:762, 831

Kennywood Park (Pittsburgh),
    4:1085
Kensington Hotel (Coney
    Island), **4**:1085
Kepler, Johannes, **4**:1150
Kettering, Charles, **3**:943–952;
    5:1412–1413
Kevlar, **1**:290, 293; **2**:351, 372;
    4:981–986, 1071, 1080
    how it works, **4**:984
Keydoozle grocery store, **5**:*1294*,
    1297
keyhole surgery, **3**:783
Khrushchev, Nikita, **3**:831
Khwarizmi, Muhammad ibn Musa
    al-, **5**:*1319*, 1320, 1323
kidney dialysis, mobile, **3**:788,
    916, 919, 920, 921
Kilby, Jack, **1**:67; **2**:341, 344, 372;
    4:965–974, 1163, 1225
Kimchee Keeper, **5**:1422
kinematoscope, **2**:502;
    4:1002–1003
kinetic energy, **3**:731
kinetograph, **2**:334, 471, 502;
    4:1002, 1003
kinetoscope, **2**:334, *470*, 471, 504
kinetoscope parlors,
    4:1002–1003, 1008
King, Augusta Ada Byron. *See*
    Byron, Augusta Ada
King, Charles Bird, **4**:1118
King, Thomas, **3**:774
Kingda Ka roller coaster, **4**:*1086*
Kingdome (sports stadium),
    1:198–199
King Kullen chain, **5**:1299
Kisk, Albert, **3**:*706*
KitchenAid, **1**:301
"kitchen debate" (1959), **3**:831
kitchen inventions, **3**:828–834,
    836
    dishwasher, **1**:295–302
    efficiency designs, **1**:195;
    3:933
    microwave oven, **1**:15, 16, 67;
    3:832–833
    Tupperware, **5**:1415–1423

kite experiments
    Bell flight, 1, 82; **1**:79
    Franklin electricity, **2**:609,
    613; **5**:1323, 1325, 1427
    Watson-Watt wireless, **5**:*1436*
Kitty Hawk (N.C.), **1**:82;
    5:1479, 1482–1483, 1484
Kitty Hawk, USS, **2**:525
Kleenex, **2**:430
    as generic term, **4**:1165
knife sharpener, **3**:847, 848
Knight, Margaret, **3**:829, 830,
    834; **4**:975–980
knitting, **1**:290
knitting machine, **1**:285–286,
    291
Koch, Robert, **3**:782, 786; **4**:1216
Kodak camera, **2**:333, 373, *458*,
    502, 506, 507; **4**:1180,
    1181, 1182
    invention of, **2**:456–457
    Polaroid suit, **4**:1226
    popularity of, **2**:504
    See also Brownie Kodak cam-
    era
Korean War, **3**:715; **4**:1078
    helicopter, **5**:1344
Kornei, Otto, **1**:247, 249
Krems, Balthasar, **3**:839
K2 (company), **4**:1168
K2Optronics, **3**:925, 929
Kuraray, **2**:430
Kurzweil, Raymond, **4**:1183, 1184
Kwolek, Stephanie, **1**:293; **2**:351,
    372; **3**:728; **4**:981–986,
    1071, 1080

labor policies, **2**:592, 597, 600;
    5:1365
Laënnec, René, **3**:781, 785
Lagoon Dipper roller coaster,
    4:1085
Lake Erie disaster (1916), **4**:1105,
    1107
Lallement, Pierre, **4**:1065–1066
Lamarr, Hedy, **2**:332, 335;
    4:987–992
lambs, cloned, **1**:239, 240

laminating, **1**:290

Lamm, Heinrich, **3**:925

landfills, **2**:517

Langen, Eugen, **4**:1193–1194, 1195, 1196

Langley, Samuel P., **2**:485; **3**:802; **5**:1479, 1480, 1481

languages, computer, **2**:370; **3**:886–887; **4**:1172
  compiler, **2**:342, 344; **3**:821–826

laparoscopy, **2**:477, 478, 479, 480, 481

laptop computer, **2**:*371*; **5**:1403, 1431

Larami Corporation, **3**:896, 897

Laserphaco Probe, **1**:72, 73

lasers, **2**:328, 370; **4**:1080, 1177, 1182, 1184; **5**:1397
  cataract surgery, **1**:69–74; **3**:783, 789; **4**:1184; **5**:1493
  compact discs, **5**:1287, 1288, 1315
  fiber-optic cable, **3**:926, 927
  invention of, **5**:1307, 1312–1315
  missile-guiding, **4**:1079
  optical disc, **5**:1289, 1315
  optical disc reuse, **4**:1202–1203
  printer, **2**:375
  spectroscopy, **5**:1313
  surgery, **1**:69, 71–74; **3**:783, 787; **5**:1493
  uses, **5**:1314–1315

latex, **3**:733

lathe, automatic, **4**:1200

Latho (sponge), **3**:793, 794

L.A. Thompson Scenic Railway Company, **4**:1081, 1082

latitude, **3**:773, 776; **5**:1407

Latrobe, Benjamin, **1**:198

Latting Observatory (N.Y.C.), **4**:1187, 1189

laughing gas (nitrous oxide), **4**:1128–1129, 1132

Launoy, M., **5**:1338–1339

Laurent, François, **4**:1097, 1100

Lauterbur, Paul, **2**:415, 416, 417

Lavoisier, Antoine, **5**:1324, 1325

law, patent. *See* patent law

Lawes, John Bennet, **2**:582, 585

lawn mowers, **1**:185–190; **2**:424
  powered, **1**:188; **4**:1183

Law of Integrated Circuits, **4**:*971*

Lawrence Berkeley National Laboratory, **3**:658, 661, 662

Lawrence Livermore National Laboratory, **5**:1387

leaded gasoline, **3**:949, 952

lead pencil, **1**:109

lead pollution, **3**:949

Leaning Tower of Pisa, **3**:665

Lebon, Philippe, **2**:433

lecithin, **3**:907–908

LED (light-emitting diode) displays, **2**:370

Led Zeppelin (music group), **4**:1235

Lee, Ezra, **1**:222–223

Lee, William, **1**:285–286, 291

Leeuwenhoek, Antoni van, **3**:801; **4**:1179

Legion of Honor Cross
  Caselli, Giovanni, **1**:280
  Cousteau, Jacques, **2**:384
  Daguerre, Louis, **2**:398
  Goodyear, Charles, **3**:725
  Jacquard, Joseph-Marie, **3**:870, 871

LEGO (company), **2**:351

Legos, **3**:917

Leibniz, Gottfried Wilhelm, **1**:42; **2**:337–338, 339, 341; **5**:1321, 1322, 1324, 1325

Lely, Peter, **3**:816

Lemelson Foundation, **2**:350–351

Lemelson-MIT Awards, **2**:349, 350–351, 352, 354
  Boyer, Herbert, **1**:154
  Cohen, Stanley, **1**:154
  Kamen, Dean, **3**:922
  Kwolek, Stephanie, **4**:986
  Lifetime Achievement, **2**:418
  Student Prize, **2**:347, 351

Lenoir, Joseph Étienne, **1**:89; **5**:1411

lenses, **4**:996, 998, 1152, 1179, 1180
  bifocal, **2**:606, 608, 613; **4**:1178–1179
  compound, **4**:1179–1180, 1181; **5**:1320–1321, 1324
  convex, **4**:994–995

Leonardo da Vinci, **3**:781, 784, 799, 862; **5**:1338

Les Paul and Mary Ford Show, The (television program), **4**:1235

letheon, **4**:1129–1130, 1132

Letterman, David, **2**:430

Leutze, Emmanuel, **4**:1046

levers and pulleys, **1**:33, 34; **2**:484; **3**:798; **5**:1319

Levi's blue jeans, **5**:1361, 1362–1364, 1365

Levi Strauss and Company, **5**:1360–1366

Levitt, Theodore, **3**:861

Leyden jar, **5**:1427, 1428

Liberkowski, Janusz, **2**:349

libraries, **2**:606

Libri, Giulio, **3**:668

license, patent vs., **4**:1220, 1221

Life (magazine), **4**:1182

life expectancy (1900–2000), **3**:789

life jackets, **2**:352

lifting action, **4**:1095, 1098

light
  bending of, **3**:924, 925, 928; **5**:1491
  electric effects, **2**:490, 548, 549, 550; **5**:1393
  electromagnetism, **4**:1016; **5**:1395
  glass-bending, **4**:995
  movement of, **2**:542
  movie projector, **4**:*1005*
  Newton experiments, **4**:*1151*, 1152
  photography, **2**:393–395, 397
  radar comparison, **5**:1439
  speed measurement, **5**:1445
  ultraviolet, **2**:516; **3**:660–662, 789
  visible, **4**:998
  wave theory, **3**:818, 849, 853–856

light (*cont.*)

  See also fiber optics; lasers; optics and vision

lightbulb, **1**:20, 197; **2**:461, 465, 468, 473, 488, 489; **4**:1021, 1022, 1026; **5**:1372

  design experiments, **3**:799

light-emitting diode (LED) displays, **2**:370

lighthouse, **1**:47

lightning, **5**:1372

lightning rod, **2**:609; **5**:1323, 1489

Lightning TRS (inline skates), **4**:1168

light pipe, **3**:925

Lilienthal, Otto, **5**:1479, 1481, 1487

limestone, **1**:194

Lincoln, Abraham, **2**:395; **4**:1049, 1220; **5**:1330, 1457

Lincoln, Patrick, **2**:495

Lincoln Memorial, **2**:*621*, 622

Lincoln Miller milling machine, **1**:309

Lindbergh, Charles, **2**:469; **3**:703, 706

Linde, Carl von, **2**:432

linen, **1**:284

Linnaeus, Carolus (Carl von Linné), **5**:1323, 1325

linotype machine, **3**:747

Linux (operating system), **2**:345, 346, 507; **4**:1227; **5**:1399–1404, 1491

lipid theory, **4**:1132

Lippershey, Hans, **3**:667–668; **4**:*993–998*, 1179, 1180, 1181; **5**:1320, 1324

liquid fuel, **4**:1192

Liquid Paper, **3**:829, 836, 860; **4**:1135–1138

liquids

  buoyancy, **1**:36–37

  repellants, **1**:166–169, 170, 293; **3**:720

  See also water

Lister, Joseph, **3**:782, 785

literacy, **3**:737, 739

Literary Piano, **5**:1332

"Little Boy" (atomic bomb), **2**:565

Liverpool and Manchester Railway, **2**:521; **5**:1354, 1355

livestock breeding, **2**:579

Livingston, Robert, **2**:*613*; **3**:650, 653, 655

Ljungquist, N. T., **3**:706

Lloyd George, David, **4**:1017

local anesthesia, **4**:1132

lock, combination, **2**:350

Lockheed Corporation, **4**:1030, 1032

locks, canal, **3**:646–647

locomotives. *See* railroads

Lodge, Sir Oliver, **1**:78; **3**:863; **4**:1018

Lodge, William, **3**:694

Log, The (electric guitar), **4**:1233, 1235

logarithms, **1**:43, 47

  table, **5**:1322, 1324

logic gates, **1**:63

London, **1**:*108*

  Great Fire (1666) and rebuilding of, **3**:818–820

  World's Fair (1851), **2**:571; **4**:1042, 1043, 1044

London, Jack, **1**:208

London Eye (Ferris wheel), **2**:575

London University, **5**:1445, 1446

Long, Crawford Williamson, **4**:1128, 1133

Long, George, **3**:694

long-distance telephone, **1**:83; **2**:328, 335

longitude, **3**:773, 774–777, 778; **5**:1407

  determining, **3**:774–775

long-playing record (LP), **2**:329, 501, 505, 507

  invention of, **3**:711, 714–715, 717–718

looms. *See* Jacquard loom; weaving

Loos, Adolf, **1**:194

Los Alamos (N. Mex.), **5**:1385, 1386, 1387

loss leader, **3**:687

Louis XVI (France), **4**:1097

Louisiana Purchase, **3**:653

Lowell, Francis Cabot, **1**:288, 293

LP. *See* long-playing record

lubrication, **5**:1409, 1412

lubricator, automatic, **4**:1049–1054

  how it works, **4**:1052

Lucent Technologies, **2**:370; **3**:860

Luddites, **1**:288; **3**:801

Ludlow, Daniel, **3**:874

Lumière, Antoine, **4**:1000, 1001, 1007–1008

Lumière, Auguste, and Louis Lumière, **2**:334, 401, 471, 502–503, 507; **4**:*999–1010*, 1181, 1182; **5**:1489

Lunar Lander (video game), **1**:231

Luther, Martin, **3**:746

luxury liners, **1**:183

lyre, **2**:*499*, 500, 505; **5**:1444

LZ-1 dirigible, **5**:*1497*, 1498, 1500

LZ-2 dirigible, **5**:1498, 1501

LZ-3 dirigible, **5**:1498, 1501

LZ-4 dirigible, **5**:*1499*, 1501

MacArthur Fellowship, **1**:100, 105

machine gun, automatic, **4**:1021–1024, *1023*, 1026, 1027, 1072

Macintosh (Mac) computer, **2**:344, 497, 498; **3**:890, 891

Macintosh, Charles, **3**:720

mackintosh raincoat, **3**:720

Mademoiselle Magazine Merit Award, **1**:70

Madersperger, Josef, **3**:839

Maffei, Scipione, **2**:387, 388

Magellan, Ferdinand, **5**:1406

Magini, Giovanni, **3**:668

Magnavox, **1**:229, 230, 231

magnetic field, **2**:415, 416

magnetic resonance imaging. *See* MRI

balloons, **5:**1496, 1497

binoculars, **4:**996, 997

cannon missile shell, **1:**110

Colt firearms, **1:**306, 307, 308, 310, *311*, 312

combat submarine, **1:**219–226

computers, **2:**339, 340, 342; **3:**822–823

dirigible, **5:**1498–1500, *1501*

Edison inventions, **2:**472–473

explosives, **4:**1070, 1157, 1164

Farnsworth inventions, **2:**554

food preservation, **1:**29, 30; **2:**347

frequency-hopping technology, **4:**987, 989–990, *991*

Global Positioning System, **3:**678, 680–681

guided missile systems, **3:**678

helicopter, **5:**1343, 1344–1345, 1413

high-tech combat uniforms, **1:**293, *294*

history of invention, **3:**798–799, 802

hydrogen bomb, **5:**1381–1389

integrated circuit, **4:**971

jet aircraft, **5:**1460, 1462–1676

Kevlar body armor, **4:**985, 986, 1071, 1080

lasers, **4:**1079, 1182

machine gun, **4:**1021–1028, 1072

nuclear devices, **2:**489–490, 557, 560, 563, 564–566, 568

plastic products, **1:**54

radar, **3:**677, 678; **5:**1309, 1310, 1397, 1437–1438, 1440–1442

railroads, **5:**1356

rifles, **5:**1449, 1455

rockets, **3:**706–708; **4:**1069, 1073, 1079, 1227

rubber uses, **3:**722

sensor-triggers, **1:**173

Teflon use, **4:**1254

telescope, **4:**997, 1179

See also gunpowder; guns; naval warfare

military bands, **5:**1303, 1306

milk pasteurization, **1:**31; **4:**1212

Milky Way, **3:**670

Mill, Henry, **5:**1332

Millennium Dome (London), **1:**199, *201*, 202

Millennium Technology Prize (Finland), **1:**105

Miller, John A., **2:**505; **4:1081–1086**

Miller, Patrick, **3:**651, 655

Miller, Phineas, **5:**1451, 1453

Miller and Baker, Inc., **4:**1083

Millikan, Robert, **2:**611

Millipede (video game), **1:**231

Miltown (drug), **3:**780

Miner's Friend (water pump), **4:**1141–1142

mines, naval, **1:**221, 222–224, 306

miniaturization, **4:**971

mini-disc player, **2:**501

minimum wage, **2:**592

mining, **1:**193, 201; **5:**1355
   dynamite, **4:**1157
   electromagnet, **5:**1370
   methods, **2:**527–530
   safety lamp, **5:**1350
   water pump, **1:**110; **4:**1140–1143, 1146; **5:**1350, 1351, 1409
   waterwheel, **4:**1237–1240, 1241

mini-submarine, **2:**381

Minix (operating system), **5:**1400–1401

Minkoff, Lawrence, **2:***413*

Minoans, **5:**1406

Minuteman (guided missile), **4:**971

Mir (space station), **4:**1202, 1203

miracle cures, **1:***305*

mirror images, **4:**1208–1209

mirrors, **3:***739*
   solar oven, **5:**1378
   telescope, **3:**669; **4:**998, 1152, 1179, 1180; **5:**1321

missiles, **1:**282; **4:**971
   defense system, **1:**65; **2:**554; **5:**1388
   integrated circuit, **4:**971

proximity fuse detonator, **4:**1077
   radar-guided, **5:**1440
   rocket-propelled, **4:**1069, 1073
   "smart," **4:**1079

"Miss Illusion" doll, **3:**793

Mississippi River steamboats, **3:**653, 655

Mistake Out Company, **4:**1135

mistakes. *See* accidents and mistakes

MIT-Lemelson Student Prize. *See* Lemelson-MIT Awards

mobile telephone. *See* cell phone

Model A (Ford car), **2:**592, 600

Model D (Deere tractor), **2:**423

Model T (Ford car), **1:**24; **2:**589, 593, 595, 596, 600, 601; **5:**1412, 1413
   price of, **2:**594, 598
   total hours to make, **2:**595

modem, **1:**85–86

molecules
   Kevlar structure, **4:**984
   protein, **1:**151, 152
   rubber, **1:**195
   spectroscopy, **5:**1309, 1311, 1312, 1313

Mollard, Jean, **2:**381

momentum, law of conservation of, **3:**856

Monadnock Building (Chicago), **2:**616, *618*

Monier, Joseph, **1:**194, 201; **4:1087–1092**

Monier system, **4:**1090–1091, *1092*

Monitor, USS, **2:**522–523, *524*, 526

Monkees, The (music group), **4:**1138

monolithic principle, **4:**968

monomers, **1:**255

Monongahela River suspension bridge, **4:**1268

Monsters, Inc. (film), **3:**889

Montagu, Mary Wortley, **3:**877

Montgolfier, Joseph-Michel, and Jacques-Étienne Montgolfier, **4:1093–1100**; **5:**1410, 1413, 1479

Montreal Protocol, **2:**516

Monument to the Great Fire
(London), **3:**819, 820

moon

gravitation, **4:**1153

telescopic observation, **4:**997,
1179

travel fantasy, **3:**699, 701–771

moon landing, **2:**348, 554, 555;
3:831; **4:**971, 1227; **5:**1413

Goddard rocketry, **3:**699, 704,
709

moons, **3:**667, 670, 672, *673*,
851, 852, 856; **4:**997

Moore, Gordon, **1:**60, 66; **2:***339*,
342, 344; **4:**971–972

Moore, Harriet, **2:***539*

Moore's law of integrated circuits,
4:*971*, 972

Morag (cloned lamb), **1:**239, 240

Morgan, Garrett, **4:**1101–1107;
5:1413

Morgan, J. P., **5:**1395

Morita, Akio, **2:**330, 336, 501;
3:829, 836; **4:**1108–1116

Morland, Samuel, **1:***213*

Morningstar Farms, **3:**941

Mörsch, E., **4:**1088

Morse, Samuel, **1:**78, 278, 279,
306; **2:**327, 334, 398, 543,
545; **3:**804; **4:**1013,
1117–1126; **5:**1428, 1446,
1447

Morse code, **1:**78; **2:**328, 462, 463;
4:1120, *1121*, 1124, *1125*

Morton, William, **3:**783, 785;
4:1127–1134

Mosaic (Web browser), **1:**101

motion, **3:**663, 668, 672–673

Newton's three laws of,
4:1149, 1152, 1153–1154

motion pictures, **2:**333–334, 401,
502–503, 505

animation, **3:**889; **4:**1103

camera, **2:**369, 401, 470, 502;
4:1005

Eastman film, **2:**458

Edison inventions, **2:**470–471,
473, 502

as industry, **4:**1008–1009

Lumières' invention of,
4:999–1010, 1181, 1182

projector, **2:**470, 471; **4:**1005

See also movie theaters;
television

motor

electric, **2:**488, 489, 540, 541,
545

Honda, **3:**805, 807, 809, 811,
812

Tesla induction, **5:***1395*, 1397

motorcycle, **2:**513; **5:***1412*

Honda, **3:**805, 807, 808–811,
812, 813, 814

invention of, **2:**403, 405; **4:**1065

milestones, **3:***810*

motorcycle helmets, **1:**293

Motorola mobile telephone,
2:358–359, 372

mouse, computer, **2:**493,
494–498; **3:**829, 830;
4:1225

how it works, 496

movable type, **2:**326, 327, 333;
3:739–740, 742, 746, 747;
5:1330, 1332

movies. *See* motion pictures

movie theaters, **2:**471; **4:**1009

first, **2:**334, 507; **4:**999, 1006

moving assembly line, **2:**589,
592–593, 594

moving stairway. *See* escalator

Mozart, Wolfgang Amadeus,
2:386, 388

Mozilla (Web browser), **1:**102

MP3 music player, **2:**337, 357

MRI (magnetic resonance imaging), **2:**351; **3:**785, 789

first machine, **2:***412, 413*, 415,
416, 418

how it works, **2:**415

invention of, **2:**409–418

MTS (mobile telephone system),
2:358

Müller, Paul, **2:**513, 582, 587

multinational companies,
4:1227–1228; **5:**1408

multiplication, **5:**1322

multitrack recordings,
4:1233–1235

mummification, **1:**284

Musher, Robert, **1:**111

music, **2:**499–501

Braille, **1:**159

digital technology, **2:**507;
5:1285–1288, 1291, 1315

downloadable files, **5:***1291*

iPod, **2:**335, 346; **3:**889, 890,
891

sales by medium, **5:***1291*

See also compact disc; long-
playing record; phonograph

musical instruments

electric guitar, **4:**1229–1236;
5:1490

piano, **2:**385–390, 500, 505

player piano, **4:**990, 991,
1230; **5:**1490

saxophone, **5:**1301–1306

Wheatstone inventions,
5:1443, 1445, 1448, 1490

musket. *See* rifle

Mussolini, Benito, **4:**988,
1017–1018

Mustang (car), **2:***602*

mutually assured destruction
(MAD), **4:**1075

Muybridge, Eadweard, **4:**1001,
1002, 1181

MySpace.com, **4:**1034

Nader, Ralph, **1:**173

Naegele, Robert, Jr.,
4:1168–1169

Nagasaki, atomic bombing of,
2:565, 567; **4:**1080, 1111;
5:1385–1386

nano assembly machine, **2:**347

nanotechnology, **1:**283, 294

Napier, John, **5:**1322, 1324

Napier's bones, **5:**1322

Napoleon I (France), **1:**29, 110;
3:648, 649, 652, 867, 868,
880; **4:**1073; **5:**1432, 1434

Napoleon III (France), **1:**280;
2:522; **3:**725; **4:**1214;
5:1305

Napoleonic Wars, **1**:29, 30, 31

NASA (National Aeronautics and Space Administration), **2**:343, 428–429; **3**:707, 709, 856, 893

National Academy of Design (N.Y.), **4**:1119

National Academy of Science, **3**:909, 911

National Air and Space Museum, **5**:1480, *1486*

National Archives, **2**:622

National Association for the Advancement of Colored People, **1**:273; **2**:446; **3**:762

National Cash Register Company, **3**:794, 945–946, 951

National Defense Research Committee, **2**:563

National Engineers Week, **2**:352

National Entrepreneur of the Year, **1**:174, 176

National Geographic Society, **1**:79
  Gold Medal, **2**:384

National Institutes of Health, **2**:415

National Inventors Council, **1**:248; **4**:990

National Inventors Hall of Fame
  Damadian, Raymond, **2**:417–418
  de Mestral, George, **2**:429–430
  Engelbart, Douglas, **2**:498
  Ginsburg, Charles, **3**:696
  Hall, Lloyd A., **3**:762
  Kilby, Jack, **4**:973, 974
  Noyce, Robert, **4**:973, 974
  Paul, Les, **4**:1235, 1236

National Medal of Merit, **1**:59

National Medal of Science
  Boyer, Herbert, **1**:154
  Cohen, Stanley, **1**:154
  Goldmark, Peter, **3**:717
  Teller, Edward, **5**:1388, 1389

National Medal of Technology
  Damadian, Raymond, **2**:417
  Engelbart, Douglas, **2**:497, 498

Kamen, Dean, **3**:922

Kwolek, Stephanie, **4**:986

Lauterbur, Paul, **2**:417

National Medals of Science for Biological Sciences, **1**:145, 154

National Safety Device Company, **4**:1103

National Science Foundation, **1**:70

National Shorthand Reporters Association, **5**:1334

National Traffic and Motor Vehicle Safety Act (1966), **1**:173–174

National Vaccine Establishment, **3**:880

Native Americans
  aspirin, **3**:780
  farming, **2**:579
  smallpox, **3**:875, 876
  transportation, **5**:1405
  wars, **1**:306, 307, 308, 311, 312
  weaving, **1**:284, 285, 291

Nativist Party, **4**:1123

natural gas, **2**:*491*

Nature (journal), **1**:237, 238, 239, 240; **2**:416; **3**:928

Nautilus (submarine), **3**:648, 649, 652, 654

Navajo code-talkers, **2**:330

Navajo weavers, **1**:284, 285

Naval Consulting Board, **2**:472

Naval Research Laboratory, **2**:371, 473

Naval Reserve, U.S., **3**:822, 824–825

naval warfare, **4**:1075–1076, 1077
  Archimedes inventions, **1**:33, 38–39
  Fulton designs, **3**:648–649, 652, 655–656
  mines, **1**:221, 222–224, 306
  submarines, **1**:219, 220–226; **3**:648–649, 652
  torpedoes, **3**:648–649, 652
  warships, **2**:520–526; **4**:1075–1076; **5**:1406

navigation, **5**:1406–1407

chronometer, **3**:773–778
  GPS, **3**:675, 678–682
  Mercator map projection, **4**:1055–1060; **5**:1407, 1409
  radar, **5**:1435, 1441–1442
  See also compass

Navstar satellite system, **3**:680, 682

Navy, U.S., **1**:58–59; **2**:521–522; **3**:908

Nazcas, **1**:79

Nazis, **2**:332; **4**:987, 988, 989; **5**:1383, 1463, 1464

negative, photographic, **2**:397, 400, 501

negative-positive photographic process, **2**:333

Nelmes, Sarah, **3**:878

Nelson, Ted, **1**:99, 100

neon lights, **5**:1395

neoprene, **1**:253, 292; **3**:722

Nepera Chemical Company, **1**:51, 55

Nerf guns, **3**:897

Nesmith, Michael, **4**:1136, 1137, 1138

Nesmith Graham, Bette, **3**:829, 836, 860; **4**:**1135–1138**

Netscape, **4**:1037
  Navigator, **1**:101, 102

Network Solutions, **4**:1036

neurology, **2**:368

neurons, **4**:1132

neutrons, **1**:16; **2**:559, 560, 562, 563

New Coke (soft drink), **4**:1249

Newcomen, Thomas, **2**:433, 485, 488, 512, 581, 584; **3**:802, 803; **4**:**1139–1148**; **5**:1350–1351, 1409, 1410

New Creations (Burbank catalog series), **1**:208–209, 210

New Orleans (steamboat), **3**:653

New Orleans jazz, **5**:*1305*, 1306

newspapers, **2**:335, 336; **3**:746
  photography, **2**:400–401
  telegraph, **4**:1125

Newton, Isaac, **1**:33; **3**:668–669, 672, 815, 818, 849, 856, 860; **4**:**1149–1156**, 1179, 1181; **5**:1321, 1325

steel (*cont.*)

    as building material, 1:*117*, 118, 193, 196, 199, 200

    electromagnet, 5:1370

    plow, 2:419–424, 511, 580, 585

    railroads, 1:193

    reinforced concrete, 4:1089, 1091

    roller coaster, 4:1085–1086

    ships, 1:183; 5:1407

    skyscraper, 2:615, 618, 621; 3:862

    strengthened, 2:539, 545

    U.S. production, 1:*111*, 116

steel-nibbed pen, 2:326

Steiner House (Vienna), 1:194

stem-cell research, 3:787

Stephenson, George, 2:485, 489, 521; 3:802, 804; 5:**1349–1358**, 1409

Stephenson, Robert, 2:521; 5:1350, 1353, 1355, 1356

Steptoe, Patrick, 2:**475–482**; 3:788

stereoisomers, 4:1208–1209

stereoscope, 5:1446

Sternbach, Leo, 3:780

stethoscope, 3:781, 790

Stevens, William, 1:114

Stockade Building System, 2:624

stocks, 2:328, 329; 4:1125

    biotechnology, 1:152

Stockton, Robert, 2:521–522

Stockton and Darlington Railway, 5:1352, 1354, 1355, 1409, 1411

stone, 1:192, 193, 200

stoves. *See* Franklin stove; ovens

Strassman, Fritz, 2:560, 561

Strategic Defense Initiative (Star Wars), 5:1388

Strauss, Levi, 1:291, 293; 3:861; 5:**1359–1366**

string instruments, 2:499, 500, 505

stroboscope, 4:1003

stroke treatment, 2:351

Strowger, Almon, 1:83, 85

Sturgeon, William, 2:544–545; 3:834; 4:1119; 5:1323, **1367–1373**

Sturm und Drang, 2:389

subatomic particles, 1:99; 2:559

submarine, 2:525, 526

    combat, 1:219–226

    detection technology, 2:554

    diesel engine, 2:438; 5:1412

    Fulton design, 3:648–649, 652, 654

    history of, 4:1076, 1077, 1078

    how it works, 1:223

    nuclear, 2:567

suburbia, 1:188

subways, 4:1260; 5:1474, 1475

Sueltz, Pat, 2:343

sugarcane, 1:127

    juice extractor, 1:109–110

Sugarman, Tracy, 4:*1050*

sulfur, 1:111, 195

sulfuric ether, 4:1129

Sullivan, William, 2:575

Sulzer Brothers, 2:438

Sumerians, 3:797, 801

sun. *See* heliocentric theory

Sunbelt, 1:265, 266

SunDome (Washington state), 1:199

Sun Microsystems, 2:343

sunscreen, 2:516

superconductivity, 1:64–65

supercooling water, 2:536

SuperCub (motorcycle), 3:811

supermarkets, 2:585, 586; 5:*1298*, 1299–1300

    frozen-food case, 1:*125*, 126

superpolyamides, 1:256

Super Skate, Inc., 4:*1166*, 1167

Super Soaker (water gun), 2:504; 3:893–898

supersonic flight, 5:1414

Supreme Court, U.S.

    color television ruling, 3:715

    patent rulings, 1:155, 279; 4:1019, 1123, 1125

surgery

    advances in, 3:783, 785, 789

    anesthesia, 4:1127, 1128–1133

    antisepsis, 3:782, 785

    artificial replacements, 3:786, 787

cataract removal, 1:69–74; 4:1184

    lasers, 1:69, 71–74; 3:783, 787; 5:1315, 1493

surrogate-mother sheep, 1:239, 240, *241*

surveying, 5:1315

suspension bridge, 1:179; 4:1263–1272; 5:1409

    how it works, 4:*1266*

sustainability, 2:518

    Lemelson-MIT award, 2:351

Swanson, Robert, 1:152, 154

sweatshops, 5:1408

sweet potatoes, 1:270, 272; 2:584

Switchback Railway (Coney Island), 4:1082

switches, electronic

    transistor, 1:63; 4:967, 972

    vacuum tube, 4:967, 971

SwitchIt Skate, 4:1169

Sylvania, 1:127; 4:991

Symantec Corp., 4:1036–1037

Symington, William, 3:651

synchrocyclotron, 2:566

synchronization, 5:1289, 1290

synthesizer, electronic, 2:507

synthetics, 2:513, 545

    drugs, 3:905, 907, 908–910, 911

    fibers, 1:290, 291–293; 2:351, 427; 3:860; 4:983

    hormones, 3:909

    See also artificial body parts; Kevlar; nylon; plastics; polymers; rubber, synthetic

Syracuse (ancient), 1:33, 34, 37, 38–40

Syracusia (ship), 1:35

Systema Saturnium (Huygens), 3:852, 857

Systers, 2:343

Syzygy. *See* Atari

Szilard, Leo, 2:564

Tab (soft drink), 4:1249

table-tennis video game, 1:227

Tacoma Building (Chicago), 2:618, 621

Taipei 101 (Taiwan), 2:622

military, **4**:1075–1076

modern suspension bridges, **4**:1263–1272

motor vehicle travel miles, **2**:*512*

waterways, **3**:645–656

wire cables, **4**:1266

See also specific types

traveling salesmen, **4**:1042, 1046

travois, **5**:*1405*

treadmill, **2**:*483, 484,* 488

Treatise on Light (Huygens), **3**:855

tree planting, **2**:514–515

trench warfare, **4**:1073

Trevithick, Richard, **5**:1351, 1409, 1410

Tri-Car, **1**:*90, 91, 92–93,* 95; **5**:1412

Trico (Tri-Continental Corporation), **1**:24, 26

tricycle, **4**:*1063,* 1068

trigonometry, **5**:1322

Trinity College (Cambridge), **1**:42; **4**:1149, 1154

triode vacuum tube, **5**:1395

Tripitaka (Buddhist sacred text), **3**:739

Trip to the Moon, A (film), **4**:1008

Triumph of the Nerds (documentary), **3**:886

Triumph phonograph, **2**:*465*

Trolleyball, **2**:450

trolley systems, **5**:1474

trombone, **5**:1305

Truman, Harry, **5**:1387

trumpet, **5**:1304

Ts'ai Lun, **2**:326, 333

tuberculosis, **3**:782, 786

Tull, Jethro, **2**:578, 579–580, 584; **3**:801, 803

Tulle-gras (bandage), **4**:1008

tungsten metal, **1**:20

tuning fork, **1**:80

tunneling shield, **1**:19

tunnels

railroad, **1**:179, 184

underwater, **1**:178, 181

Tupper, Earl, **2**:513; **3**:829, 830, 835; **5**:**1415–1424**

Tupper Home Parties, **5**:1421–1422

Tupperware, **2**:513; **3**:728, 830, 835; **5**:1415–1423

Tupperware party, **5**:1420, 1421–1423

turbine

jet engine, **5**:1461, 1463

steam, **2**:488, 489; **5**:1396

water, **4**:1238–1240, 1241

Turbosails, **2**:383

Turri, Pellegrino, **4**:1183; **5**:1332

Turtle (first combat submarine), **1**:*220, 221–223, 224, 225–226;* **4**:1076, 1077

Tuskegee Institute, **1**:27, 269, 271, 273–274; **2**:458; **3**:894

Tuve, Merle, **5**:1437

Twain, Mark, **1**:15, 20; **4**:1228

two-stroke engine, **1**:90; **2**:439

Tymshare, Inc., **2**:497

Tyndall, John, **3**:925

typewriter, **1**:161; **2**:327, 328, 334, 338; **4**:1183; **5**:1447

for blind people, **5**:1332

correction fluid, **3**:829, 836, 860

Henry inventions for, **3**:793–794, 795

how it works, **5**:1331

invention of, **5**:1329–1336

Liquid Paper, **4**:1135–1138

patent rights, **4**:1225

Shales invention, **3**:829, 834

shift key, **5**:1333, 1335

Typographer machine, **5**:1332

ultrasound cataract surgery, **1**:72, 73

ultraviolet (UV) light, **2**:516; **3**:789

water purification, **3**:657–662, 789

umbrella, snap-on cloth cover, **3**:792–793, 794

Undersea World of Jacques Cousteau, The (documentary), **2**:380

underwater exploration, **2**:379–382; **5**:1325, 1326, 1327, 1328

underwater tunnel, **1**:178, 181

undulatory theory, **3**:855

unicycles, **4**:*1062*

Uniform Resource Locator, **1**:100, 102

Union Carbide and Carbon Corporation, **1**:54, 55

Union Pacific Railroad, **5**:1357

unions, **2**:600

Union Tank Car Company, **1**:198

Unisys Corporation, **1**:217

United Aircraft Corporation, **5**:1341

United Nations, **1**:*105,* 162

Food and Agriculture Organization, **3**:762

UNIVAC-1 computer, **3**:823, 824

universal joint, **3**:818

University of California, Berkeley, **2**:495; **3**:658, 661, 884, 888; **5**:1313

Strauss-funded scholarships, **5**:1362

University of Chicago, **2**:562, 564, 565; **4**:1073

University of Illinois, **1**:59, 64–65, 101, 252, 256

Assembly Hall dome, **1**:199

University of Minnesota, **1**:59, 140, 145

University of Pennsylvania, **2**:339, 340, 608

Medical School, **1**:148, 153

University of Vienna, **3**:906–907, 908

UNIX (operating system), **2**:370; **5**:1400, 1401, 1403

Unsafe at Any Speed (Nader), **1**:173

"ups-and-downs," **2**:571, *572*

uranium, **2**:559, 560, 562, 563; **5**:1383, 1384

Uranus (planet), **4**:1180

Urban VIII, Pope, **3**:671

urbanization, **2**:509–511, 512

Urban League, **3**:762

URL (Uniform Resource Locator), **1**:100, 102

U.S. Electric Lighting Company, **4**:1022, 1024

utopianism, **3**:688

UV Waterworks system, **3**:657, 658, 659, 660–661, 789

uWink.com, **1**:231, 233

vaccine and vaccination, **1**:18; **3**:799, 879
  how it works, **4**:*1215*
  Pasteur, **4**:1207, 1209, 1214–1216, 1217
  smallpox, **3**:782, 785, 873–882

vacuum canning, **3**:899–904

vacuum cleaner, **2**:488; **3**:832, 834, 836
  bagless, **2**:449–452

vacuum tube, **1**:60, 61; **2**:339–341, 345; **4**:967, 971; **5**:1395

Vail, Alfred, **4**:1120, 1125

Vanderbank, John, **4**:*1150*

variolation, **3**:876, 877–878, 879

varnish, **1**:53

Vatican, **4**:1018, 1189

VCR. *See* videocassette recorder

vegetarianism, **3**:938

Velcro, **1**:16, 19, 293; **2**:425–430; **3**:799

Velcro Industries B.V., **2**:430

Velcro S.A., **2**:427, 429

vellum, **2**:326; **3**:743

Velo (Benz car), **1**:94

velocipede, **4**:1062, 1063–1064, 1065, 1066, 1067

Velox, **1**:51, 55

Verne, Jules, **5**:1338

Vesalius, Andreas, **3**:781, 784

VHNS (Vertical Helical Scan), **4**:1114

Victoria, Queen (Britain), **1**:78, 109–110; **2**:544; **4**:979, 1043

Victor records, **2**:329. *See also* RCA

Victory (ship), **2**:520, 522

video camera, **4**:1114

videocassette recorder (Betamax), **4**:1114, 1115

videocassette recorder (VCR), **2**:330; **3**:697
  EVR predecessor, **3**:717

videoconferencing, **2**:495–496

video digital recording. *See* DVD

video games, **1**:227–234; **2**:349, 506, 508; **3**:689

videotape recorder (VTR), **2**:329, 330, 335, 505, 550; **3**:691–698, 829, 836
  as DVD replacement, **5**:1291

Vietnam War, **1**:176; **4**:1078

Viking probes, **3**:*709*, 710

Vikings, **4**:1075

Viktoria (Benz four-wheeled car), **1**:*92*, 93

Vineland Police Department (N.J.), **1**:138

vinyl records. *See* long-playing record

violin, **2**:500, 506

Virchow, Rudolf, **3**:781, 786

Virginia, CSS. *See* Merrimack

virtual wars, **4**:1079

viruses (computer), **4**:1029, 1032–1033, 1036
  effect of, *1034*

viruses (organisms), **1**:149; **3**:782, 875, 882; **4**:1209
  vaccines, **4**:*1215*

VirusScan (software), **4**:1032, 1037

Viscoloid, **5**:1417

vision. *See* blindness; lenses; optics and vision

vitascope, **2**:471

Vitruvius, **2**:484

VLSI (integrated circuit), **2**:345

Volksempfanger, **4**:1079

Volkswagen, **2**:595

volt (electric unit), **5**:1434

Volta, Alessandro, **2**:486, 488, 539, 611; **5**:1323, 1325, **1425–1434**

Volta prize, **1**:78

volume computation, **1**:40

V-1 and V-2 rockets, **3**:677, 708; **4**:1073

VS-300 helicopter, **5**:1344

VTR. *See* videotape recorder

Vukic, Ivan Blaz Lupis, **2**:523–524

vulcanized rubber, **1**:195, 200; **5**:1410, 1411
  as accidental discovery, **1**:16, 17–18, 20; **3**:723
  development of, **3**:719–728
  how it works, **3**:727; **4**:1225

wagons, **5**:1409

Walker, Samuel, **1**:307

Walker Colt revolver, **1**:307

walkie-talkie radio, **2**:358

Walkman, **2**:330, 336, 501, 508; **3**:829, 834, 836; **4**:1108, 1112–1116
  naming of, **4**:1113

Wal-Mart, **5**:1403

Waltham (Mass.), **1**:288

Waltham Engineering, **1**:172, 174

Wan-Hoo crater (moon), **3**:707

Wan Hu, **3**:707

war. *See* military and weaponry; *specific wars*

Waring, Fred, **4**:1232

Warner Communications, **1**:231, 232

War of 1812, **3**:655

War of the Worlds (Wells), **3**:699

warps, **1**:290

Warren, John Collins, **4**:1129

warships, **4**:1075–1076; **5**:1406
  ironclads, **2**:522–524; **4**:1076
  screw-propelled, **2**:521–526
  See also submarine

washing machine, **2**:452; **3**:831, 834

Washington, Booker T., **1**:269, 270

Washington, George, 656; **1**:222; **3**:648

Washington Monument, **4**:1189

waste materials, **2**:513–514, 517, 567

watches
  balance springs, **3**:817
  digital, **1**:67, 246; **2**:342; **3**:885
  GPS receivers, **3**:682

PHOTOGRAPHIC CREDITS

Cover photos (clockwise from left): David Greedy/Getty Images; John T. Barr/Getty Images; Jupiter Images; China Photos/Getty Images. **Associated Press:** 1401; **Corbis:** Bettman 1294; Corbis 1471; W. Dickson 1490; **Getty Images:** AFP/AFP 1285, 1289; Herbert Barraud/Hulton Archive 1392; Boyer/Roger Viollet 1311; Central Press/Hulton Archive 1460; China Photos/Getty Images News 1367, 1371; Alfred Eisenstaedt/Time & Life Pictures 1308, 1389; Rob Elliot/AFP 1365; Express/Hulton Archive 1464; Fox Photos/Hulton Archive 1385, 1436; Sean Gallup/Getty Images News 1459, 1467; Getty Images/Getty Images News 1321, 1441; Allan Grant/Time & Life Pictures 1296; Chris Graythen/Getty Images Entertainment 1305; Charles Hewitt/Hulton Archive 1463; George H. H. Huey 1407; Hulton Archive/Hulton Archive 1340, 1354, 1444, 1497; Nicholas Kamm/AFP 1469, 1474; Koichi Kamoshida/Getty Images News 1433; Keane Collection/Hulton Archive 1394, 1470; Raymond Kleboe/Hulton Archive 1465; John Kobal Foundation/Hulton Archive 1363; Dmitri Kessel/Time & Life Pictures 1334; John G. Mabanglo/AFP 1400; Mansell/Time & Life Pictures 1352; MPI/Hulton Archive 1454, 1456; Daniel Pepper/Getty Images News 1452; Per-Anders Pettersson/Getty Images News 1399, 1403; Frank Scherschel/Time & Life Pictures 1338; Sam Shere/Hulton Archive 1501; Ted Streshinsky/Time & Life Pictures 1310; Justin Sullivan/Getty Images News 1359, 1360, 1477, 1487; Time & Life Pictures/Time & Life Pictures 1302, 1307, 1315, 1381, 1386, 1387, 1406; Sam Yeh/AFP 1304; Underwood and Underwood/Time & Life Pictures 1496; USAF/Getty Images Sport 1435, 1439; **iStockphoto:** bluestocking 1319; buzbuzzer 1349, 1357; fotofrog 1317; MiguelAngeloSilva 1327; resonants 1286; subman 1346 **Jupiter Images:** 1293, 1298, 1318, 1329, 1331, 1335, 1337, 1342, 1350, 1353, 1356, 1425, 1426, 1427, 1429, 1432, 1445, 1449, 1450, 1451, 1486, 1500; **Library of Congress:** 1375, 1376, 1478, 1485; **NASA:** 1481, 1482; **Science Museum/Science and Society Picture Library:** 1330, 1339, 1351, 1368, 1391, 1395, 1405, 1412, 1431, 1443, 1446; **Smithsonian Institution:** 1416, 1419, 1421; **Tupperware Brands Corporation:** 1415, 1424; **United States Patent Office:** 1287, 1295, 1361, 1377, 1418, 1473, 1484, 1489.